Heterobimetallic [2.2]Paracyclophane Complexes and Their Application in Photoredox Catalysis

Zur Erlangung des akademischen Grades eines/einer

DOKTORS/DOKTORIN DER NATURWISSENSCHAFTEN

(Dr. rer. nat.)

von der KIT-Fakultät für Chemie und Biowissenschaften

des Karlsruher Instituts für Technologie (KIT)

vorgelegte

DISSERTATION

von

M. Sc. Daniel Maximilian Knoll

aus Kassel

Dekan: Prof. Dr. Manfred Wilhelm

Referent: Prof. Dr. Stefan Bräse

Korreferent: Prof. Dr. Peter Roesky

Band 85
Beiträge zur organischen Synthese
Hrsg.: Stefan Bräse

Prof. Dr. Stefan Bräse
Institut für Organische Chemie
Karlsruher Institut für Technologie (KIT)
Fritz-Haber-Weg 6
D-76131 Karlsruhe

Bibliographic information published by the Deutsche Nationalbibliothek

The Deutsche Nationalbibliothek lists this publication in the Deutsche Nationalbibliografie; detailed bibliographic data are available in the Internet at http://dnb.d-nb.de

ISBN 978-3-8325-5071-4
ISSN 1862-5681

Logos Verlag Berlin GmbH
Comeniushof, Gubener Str. 47,
10243 Berlin
Tel.: +49 030 42 85 10 90
Fax: +49 030 42 85 10 92
INTERNET: http://www.logos-verlag.de

Für Mama, Papa und Melanie.

"Errare humanum est, sed perseverare diabolicum."

– Seneca

HONESTY DECLARATION

This work was carried out from 2016 December 1st through 2019 November 6th at the Institute of Organic Chemistry, Faculty of Chemistry and Biosciences at the Karlsruhe Institute of Technology (KIT) under the supervision of Prof. Dr. Stefan Bräse.

Die vorliegende Arbeit wurde im Zeitraum vom 1. Dezember 2016 bis 6. November 2019 am Institut für Organische Chemie (IOC) der Fakultät für Chemie und Biowissenschaften am Karlsruher Institut für Technologie (KIT) unter der Leitung von Prof. Dr. Stefan Bräse angefertigt.

Hiermit versichere ich, Daniel Maximilian Knoll, die vorliegende Arbeit selbstständig verfasst und keine anderen als die angegebenen Hilfsmittel verwendet sowie Zitate kenntlich gemacht zu haben. Die Dissertation wurde bisher an keiner anderen Hochschule oder Universität eingereicht. Die „Regeln zur Sicherung guter wissenschaftlicher Praxis am Karlsruher Institut für Technologie (KIT)" wurden beachtet.

Hereby I, Daniel Maximilian Knoll, declare that I completed the work independently, without any improper help and that all material published by others is cited properly. This thesis has not been submitted to any other university before.

German Title of this Thesis

Heterobimetallische Komplexe des [2.2]Paracyclophans und deren Anwendung in der Photoredoxkatalyse

ABSTRACT

The most important goal of current chemistry research is to provide green and sustainable routes to compounds of interest. One way of addressing this is the use of abundant and inexpensive sources of energy to drive reactions, with the prime example being visible light in photoredox catalysis. One recent promising approach is the use of heterobimetallic catalysts where two metals work in a cooperative fashion to achieve the desired transformation. However, very little is known about the exact mechanism of cooperativity. This is due to a lack of heterobimetallic compounds that can be fine-tuned to obtain very specific answers regarding for example spatial design of the catalyst in question.

In this work, [2.2]paracyclophane (PCP) is presented as a new platform on which to build distance-variable heterobimetallic complexes. PCP has been described as a "super-atom" molecule for its unique ability to hold up to 16 substituents in a precise spatial relationship to each other. By using PCP as a platform, it is shown that a range of defined heterobimetallic complexes can be designed and prepared (Figure 1.1). The methods necessary for the synthetic transformations are developed and investigated for their broader synthetic applicability.

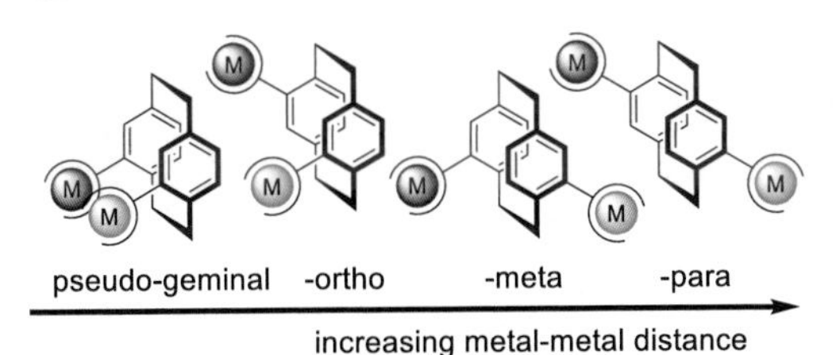

Figure 1.1. Distance modulation of two catalytic metal centers.

To demonstrate the potential of these complexes as catalysts, Au/Ru heterobimetallic complexes are evaluated regarding their performance in a dual photoredox catalytic arylative Meyer-Schuster rearrangement reaction. This reaction provides a very convenient and sustainable access α-arylated enones, an important building blocks for pharmaceutical relevant compounds.

Kurzfassung

Eines der wichtigsten Ziele der modernen Chemie ist die Entdeckung grüner und nachhaltiger Syntheserouten. Eine Möglichkeit dieses Ziel zu erreichen besteht in der Verwendung günstiger und gut verfügbarer Energiequellen, um eine Reaktion voranzutreiben. Hierbei stellt sichtbares Licht ein Paradebeispiel dar. In den letzten Jahren haben sich dafür heterobimetallische Katalysatoren, in denen die beiden enthaltenen Metallatome kooperativ miteinander arbeiten, als besonders geeignet erwiesen. Jedoch ist bis jetzt sehr wenig über den exakten Mechanismus der Kooperativität bekannt. Dies liegt nicht zuletzt an der geringen Verfügbarkeit und Einstellbarkeit der bekannten Heterobimetallkomplexe, was zur Beantwortung spezifischer Fragestellungen hinsichtlich der räumlichen Anordnung der Metallatome notwendig wäre.

In dieser Arbeit wird das [2.2]Paracyclophan (PCP) als neuartige Plattform vorgestellt, auf der distanzvariable Heterobimetallkomplexe aufgebaut werden können. PCP wurde als Superatom bezeichnet, da es die einzigartige Eigenschaft besitzt bis zu 16 Substituenten in einer präzisen räumlichen Struktur zueinander zu fixieren. Mithilfe des PCP wird gezeigt, dass eine Reihe von definierten Heterobimetallkomplexen entworfen und synthetisiert werden kann. Die dazu notwendigen Methoden für die synthetischen Transformationen werden entwickelt und auf ihre generelle synthetische Anwendbarkeit hin untersucht.

Um das Potential der hergestellten Komplexe als Katalysatoren zu demonstrieren, werden die Au/Ru Heterobimetallkomplexe hinsichtlich ihrer Leistungsfähigkeit in einer dual photoredoxkatalysierten arylierenden Meyer-Schuster-Umlagerungsreaktion getestet. Diese Reaktion bietet einen einfachen und nachhaltigen Zugang zu α-arylierten Enonen, die ein wichtiger Baustein für die Synthese von pharmazeutisch relevanten Verbindungen sind.

CONTENTS

1 INTRODUCTION

1.1 [2.2]Paracyclophane

1.1.1 Introduction

The backbone and molecule of interest in this work is [2.2]paracyclophane (PCP, **1**), a molecule displaying an intriguing shape and properties. Its structure (Figure 1.1) features two benzene rings (decks) that are cofacial-stacked and held in this arrangement by two ethyl bridges attached at the *para* positions of the benzene rings.

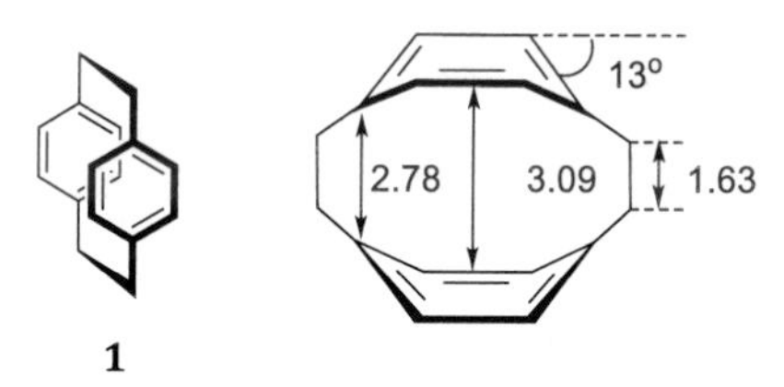

Figure 1.1. Structure of PCP. Bond lengths and distances are given in Å.

These short bridges force the decks closer together than would be energetically favorable, thereby inducing a strain. The distance between the benzene rings (3.09 Å) is shorter than the van-der-Waals distance of the layers in graphene (3.40 Å).[1] Two phenomena arise from this strain: (i) a "bent and battered", boat-like form with the bridgehead carbons being shifted out-of-plane is adopted by the benzene rings, caused by repulsive forces between them,[2] (ii) transannular through-space electronic communication between the decks is enabled through the close proximity and thus overlap of the π-systems of the benzene rings.[3]

The former phenomenon (i) leads to a less-than-aromatic character of the benzene rings. This can be seen spectroscopically, e.g. in the ^1H NMR spectrum of PCP, where the aromatic proton peaks are shifted around 1 to 1.5 ppm upfield. It also affects synthetic transformations, exhibiting chemical behavior that sometimes differs significantly from the *p*-xylene (the "monomer" of PCP) chemistry.

The latter phenomenon (ii) has implications for the functionalization of the molecule. For example lithiation of one deck increases the electron density not only on one but also on the other deck, a lithiation of the not-yet-lithiated deck becomes energetically less favorable.[4] Furthermore, the electronic communication between the decks was reported to be applied in molecular junctions[5] and molecular wires.[6,7]

Both the unusual shape and electronic situation of the benzene rings are accompanied by the sterical encumbrance that is brought about by the *para*-disubstitution and additional shrouding of one face of each benzene ring by the other deck. Combined, these circumstances lead to chemical behavior that is markedly different from isolated *p*-xylene or even benzene, including sluggish or unexpected reactivity. One example, the cross-coupling chemistry of PCP, is discussed in detail in section 3.3.

1.1.1.1 Discovery and Nomenclature

The first discovery of PCP was reported in 1949 by Brown and Farthing, when they were analyzing the polymerization products of *p*-xylene.[8] Since then PCP has come a long way from a "lab curiosity" to a large array of applications in asymmetric catalysis, material sciences, supramolecular chemistry and medicine.[1]

The nomenclature of these molecules following IUPAC rules can be quite laborious, thus a new nomenclature system was invoked. According to Vögtle *et al.*, the term "phane" describes a structure that contains at least one aromatic or **ph**enyl ring bridged by an alk**ane.** If the aromatic ring is indeed a benzene derivative, the class if cyclophanes is described. The number of atoms in the bridging chains is given by an integer for each bridge separated by a period in square brackets in front of the name. Finally, in case of cyclophanes the relative orientation of the bridgeheads on the benzene is given by the common prefixes *ortho, meta, and para*.[9]

1.1.1.2 Chirality

Another remarkable feature of PCP arises from the impeded rotation of the rings along their bridge-to-bridge axis. Introduction of a substituent other than hydrogen at any of the carbon centers renders these molecules chiral. If the bridges are substituted, common central chirality is observed, but when the substituent is located on one of the decks, the molecule becomes planar chiral (Figure 1.2). Interestingly, most compounds featuring planar chirality bear the PCP scaffold in their structure.[10]

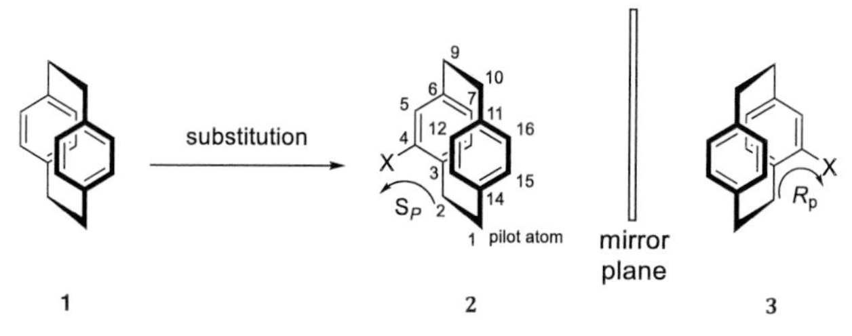

Figure 1.2. Planar chirality of PCP.

Remarkably, PCPs configurational stability tolerates temperatures of up to 200 °C before racemization is observed. The racemization occurs through homolytic bond cleavage between the bridge carbon atoms, thus freeing PCP of its hindered rotation about the decks and yielding a statistical distribution of both recombination products (Scheme 1.1). This phenomenon was exploited in a polymerization technique known as the Gorham process (see section 1.1.2.1.).[11]

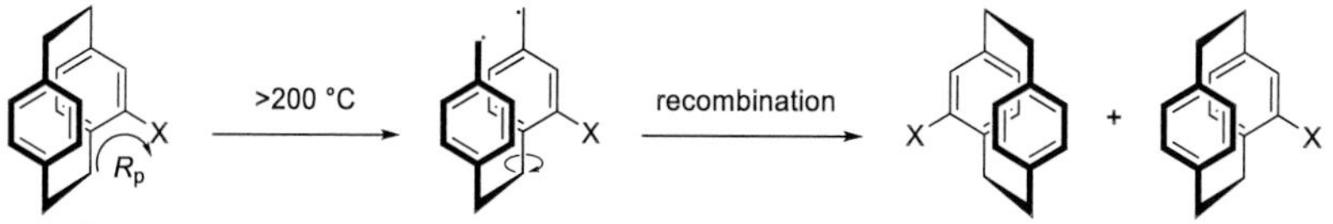

Scheme 1.1. Racemization of PCP above 200 °C.

If two substituents are located on the decks, a total of 7 regioisomers are observed (Figure 1.3). The prefix "pseudo" is used when the substituents are on differing decks. For regioisomers with higher symmetry, the substituents must differ from each other (=heterodisubstitution) to enable planar chirality.

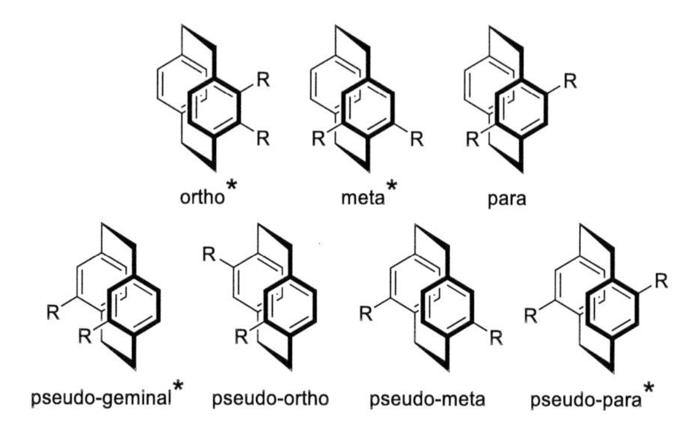

Figure 1.3. Regioisomers of deck disubstituted PCP. Compounds marked with an asterisk are only chiral if heterodisubstituted.

1.1.2 Applications of PCP

1.1.2.1 The Gorham Process

Already in 1947, Szwarc noted a brown residue forming in the cooler zones of his pyrolysis apparatus that he used to investigate hydrocarbons and their pyrolytic behavior. He disassembled the apparatus to find that the formed residue could be peeled off much like the "skin of a snake" and realized that a new polymer had been formed. He accidentally made the first vapor-deposited polymer. The process itself was significantly improved by Gorham at Union Carbide, finding optimized conditions at 550 °C and vacuum below 1 Torr.[12]

In the Gorham process, both ethylene bridges are cleaved to produce an intermediate *para*-xylylene that polymerizes on cool surfaces to generate parylene (Scheme 1.2). The starting material for all parylenes is PCP with various substituents that end up in the finalized polymer, thus giving the polymer tailor-made properties. Not surprisingly, this versatile process found numerous applications such as implantable electronics[13], biological,[14] microfluidical[15] and optical materials.[16] Through the vapor-deposition process, nearly surfaces of every shape and property can be coated evenly with a pinhole-free surface finish. The parylene coating shows useful properties like chemical inertness, temperature stability and biocompatibility.[17]

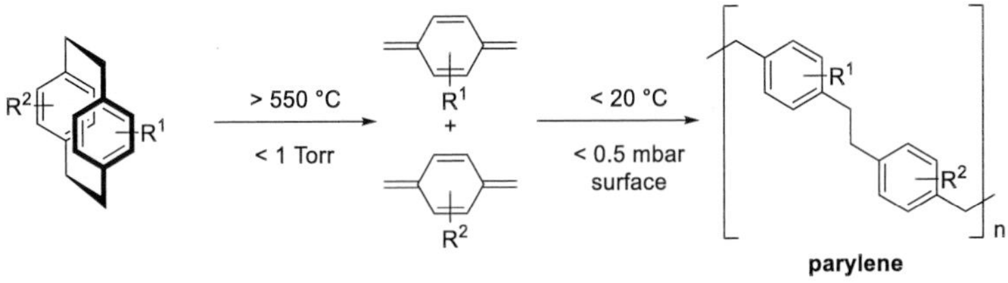

> 550 °C
< 1 Torr
< 20 °C
< 0.5 mbar surface
parylene

Scheme 1.2. The Gorham Process.

1.1.2.2 Asymmetric Catalysis

PCP has been named the "sister molecule of sandwich compounds like ferrocene"[10] due to the strikingly similar structure. Like ferrocene, which is by far the most widely used planar chiral ligand system, PCP has been employed for a wide range of asymmetric catalytic transformations. The sheer amount of data published on asymmetric catalysis with PCP cannot be adequately summarized in this work, thus the reader is referred to overview articles on the topic.[1,18–22]

Exclusively making use of planar chirality, PhanePhos (Figure 1.4) has been used in rhodium and ruthenium-catalyzed stereoselective hydrogenation reactions, similar to its ferrocene analogue dppf (**5**).[23,24]

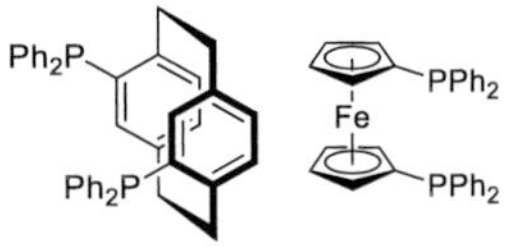

Figure 1.4. Phanephos, a planar chiral ligand used in stereoselective hydrogenation and 1,1'-bis(diphenylphosphino)ferrocene (dppf).

However, combined results of researchers looking into paracyclophanes show that planar chiral ligands can lack the amount of enantioinduction that their central chiral counterparts offer.[25] Thus, efforts have been made to combine planar chirality with central chirality. However, a unified concept on how to design planar/central chiral catalysts has not been identified to date and more research is necessary.[26]

Some *N,S*- and rather exotic *N,Se*-planar chiral PCPs were successfully employed by Hou *et al.* in a palladium-catalyzed allylic alkylation reaction at ambient conditions (Scheme 1.3).[27] They noted the significant difference in enantioselectivity if the coordinating atom was located either at the benzene ring in (*S*, *Rp*)-8 or at the benzylic position in (*R*, *S*, *Rp*)-9, with the latter performing markedly better. The same result was observed for the *Se*-analogue.

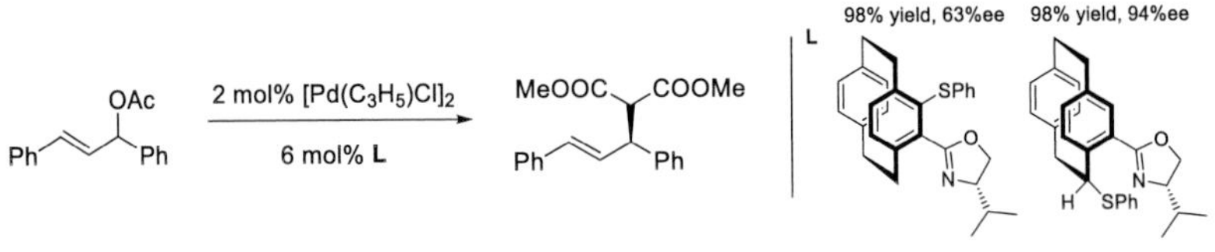

2 mol% $[Pd(C_3H_5)Cl]_2$, 6 mol% **L**

6 → (*S*)-**7**

L: (*S*, R_p)-**8**: 98% yield, 63%ee; (*R*, *S*, R_p)-**9**: 98% yield, 94%ee

Scheme 1.3. Palladium-catalyzed allylic alkylation with two different PCP ligands.

However, not all PCP catalysts are ligands for metal catalysis. An *et al.* recently reported a metal-free enantioselective silylation of aromatic aldehydes by an *N*-heterocyclic carbene PCP catalyst (Scheme 1.4).[28] The obtained chiral α-hydroxysilanes were accessible in moderate to excellent yields and enantioselectivities. With this method, the use of air and moisture sensitive copper or ruthenium catalysts can be circumvented.

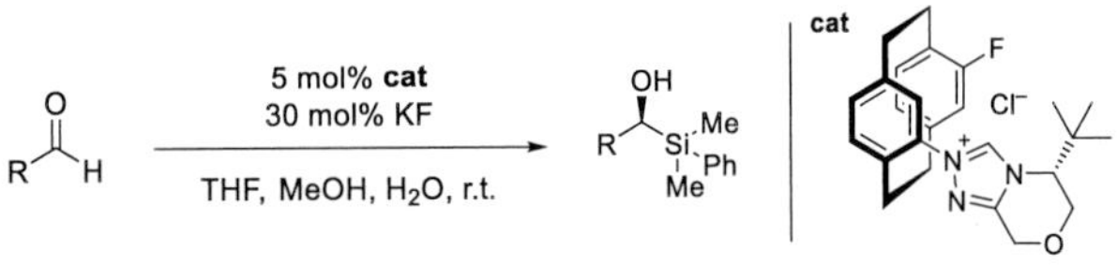

5 mol% **cat**, 30 mol% KF, THF, MeOH, H_2O, r.t.

10 → **11**

cat: (*S*, S_p)-**12**

Scheme 1.4. Metal-free enantioselective silylation of aromatic aldehydes by an *N*-heterocyclic carbene PCP catalyst

These examples demonstrate the very wide field of PCP ligands and catalysts that have been applied for and probably will continue to remain.

1.1.2.3 PCP as Spacer and Linker

Due to PCP's rigidity and configurational stability, all of its 32 atoms are fixed in their relative orientation. This has led to the term "hexadecavalent superatom" as it can be regarded as one solid sphere with up to 16 attachable functional groups.[10] Furthermore, the rigidity of the PCP core is transferred to the substituents that accordingly also inhabit precise positions in three-dimensional space. Thus, PCP presents itself as a way to tie up to 16 elements together to study effects arising from the spatial relationship between those elements.

This has predominantly been used for the preparation of chiroptical structures making use of the spatial organization provided by the PCP core.[1] One recent example using tetrasubstituted PCP was published by Chujo *et al.*[29] Here, a propeller-shaped molecule is generated from PCP and its structure allows for circularly polarized luminescence to occur (Figure 1.5). The chirality of PCP is directly transferred into the chirality of the emitted photons.

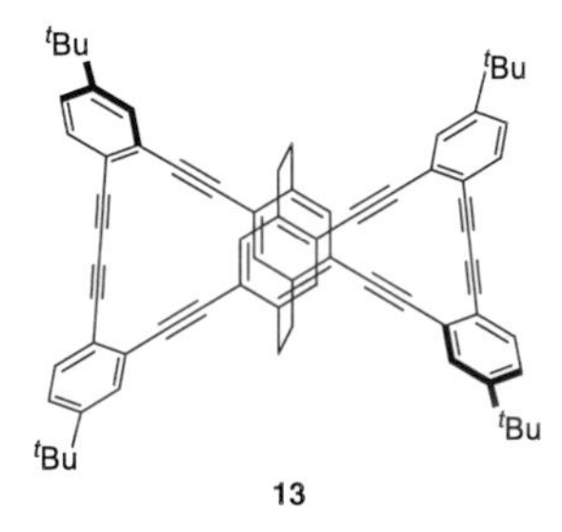

Figure 1.5. Propeller-shaped **13** emits circularly polarized light when excited.

This ability of PCP to fix positions of substituents in space lends itself conveniently to the aim of this work, the preparation of fixed-distance heterobimetallic complexes. To avoid problems of steric repulsion, a maximum number of two substituents was planned to be introduced to the PCP core. Additionally, while the position of the substituent itself is fixed, free rotation is still possible. A complete fixation of the metal bound to the binding site can be achieved by steric bulk of the binding site to avoid free rotation. Another possibility is using a cyclometallation motif that binds back to the PCP, thus attaching the metal atom directly in one of the fixed positions on the PCP core.

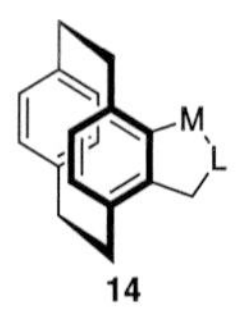

Figure 1.6. Binding site with cyclometallated motif **14** fixates the position of the metal center to the PCP core.

1.1.3 Pyridyl PCP

One of the targeted coordination motifs in this work is pyridyl PCP **15**, the PCP analogue of 2-phenylpyridine **16**, a widespread structural motif in coordination chemistry (Scheme 1.5). The latter has seen a surge of interest for its application in highly luminescent cyclometallated Ir(III) and Pt(II) complexes that are used in organic light emitting diodes (OLEDs).[30] Additionally, cyclometallated Pt(II) or Ru(II) complexes bearing 2-phenylpyridine (**16**) as ligand have shown potential as an alternative to well-established cis-platinum compounds used as anti-tumor medication or due to their luminescence as imaging agents for cancer cells.[31–33] The cyclometallated Pt(II) or Ru(II) complexes show a dramatically increased anticancer activity compared to their non-cyclometallated analogues.[34]

structural analogue to

15 **16**

Scheme 1.5. 4-(2'pyridyl)[2.2]paracyclophane as structural analogue to 2-phenylpyridine.

Previous work towards the synthesis of PCP pyridyl **15** can be classified in *de novo* syntheses, traditional cross-coupling approaches and cross-coupling of pyridine N-oxides.

Hopf *et al.* reported a *de novo* synthesis of **15** starting from 4-acetyl[2.2]paracyclophane (**17**). After refluxing with *N*-methylhydroxylamine, nitrone **18** is formed. Further refluxing the obtained nitrone **18** with 1,2-dibenzoylethene led to the unexpected product **19** in good yield (Scheme 1.6).[35] However, this synthesis route provides neither the flexibility or modularity associated with modern cross-coupling reactions.

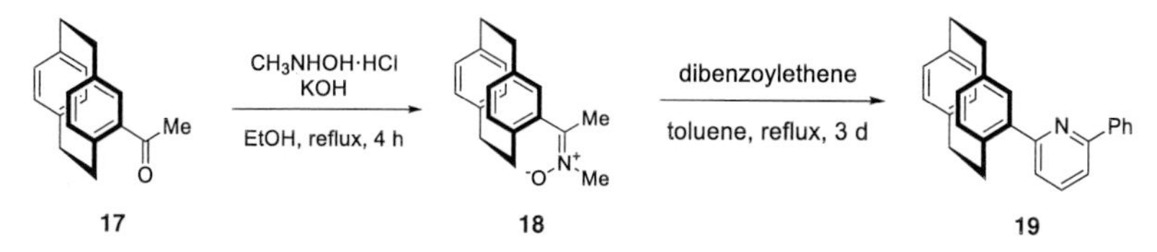

Scheme 1.6. *De novo* synthesis of pyridyl PCP **19**.

A more traditional Stille cross-coupling approach towards pyridyl PCP is described in more detail in section 3.3.1.4. This synthesis pathway using stannylated pyridines or stannylated PCP is significantly better suited for a modular synthesis of pyridyl PCPs. However, the toxicity and complexity of handling of organotin reagents demand for a more benign and green approach.

In 2010 Fulton *et al.* reported a direct cross-coupling of PCP bromide with pyridine N-oxides[36] inspired by the chemistry of Fagnou *et al.*[37] The reaction is modular with a wide substrate scope of differently functionalized pyridines using inexpensive $Pd(OAc)_2$ (Scheme 1.7). However, the yields are only moderate and suffer drastically upon the introduction of more substituents on the side of the PCP. Since the molecules sought after in this work are heterodisubstituted PCPs, another type of reaction could offer easier synthetic access.

5 mol% $Pd(OAc)_2$
t-$Bu_3P{\cdot}HBF_4$
K_2CO_3
toluene, reflux, 15 h
Br
+
O-
N+
N
51 **20** **15** (66%)

Scheme 1.7. Direct cross-coupling of pyridine N-oxide **20** with PCP bromide **51** to furnish pyridyl PCP **15**.

It should be noted that the 2-phenylpyridine motif would alternatively be accessible by arylation of a pyridinophane leading to phenylpyridinophane **21** (Scheme 1.8). However, these molecules are notoriously difficult to make to be considered a valid pathway.[38,39]

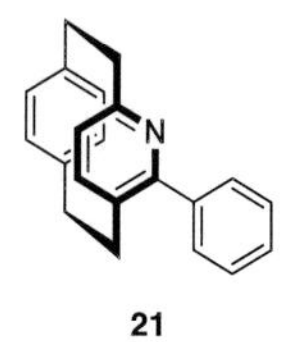

Scheme 1.8. Arylated pyridinophane **21** as an alternative 2-phenylpyridine containing structure.

1.2 Photoredox Catalysis

1.2.1 Introduction

Ultimately, the sun is the only influx of energy for all processes happening on earth with only insignificant amounts of radiation heat from the earth's inside. For this reason, scientists have dreamed for centuries to directly convert sunlight to more useful forms of energy. While this has been achieved with reasonable success for the conversion of light energy to electrical energy (solar cells), the conversion into chemical energy remains challenging. Nature itself has perfected the latter in photosynthesis, but chemists only started to grasp how to harness the sun's free and abundant energy to make new compounds.

1.2.1.1 Discovery

Only recently, Stephenson, Yoon and MacMillan simultaneously published a new concept that would change the way scientists thought about driving reactions with light.[40–42] While light was previously only really used in the form of high-energy bond-breaking UV light, this new field makes use of visible light, utilizing ordinary light bulbs or even low-cost and energy-efficient LEDs. In the last decade, since its first broad investigation, the number of publications has reached over 40,000 according to a *Google Scholar* search for "photoredox catalysis". This goes to show the importance and continuing interest of researchers in the field.

1.2.1.2 Applications

The advent of photoredox catalysis led to the review and improvement of known synthetic transformation. Benefits range from milder and greener conditions to catalytic transformations that previously required stoichiometric amounts of reagents. Some examples for this include the reduction of azides to amines with a ruthenium catalyst and visible light (Scheme 1.9)[43] or the photocatalytic version of the Pschorr reaction (Scheme 1.10).[44]

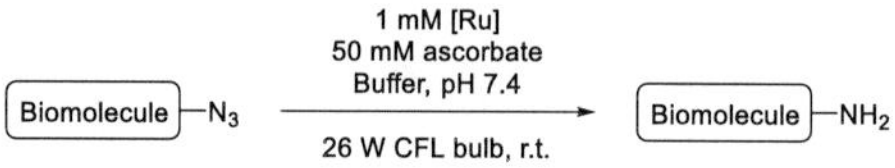

Biomolecule—N_3 → 1 mM [Ru], 50 mM ascorbate, Buffer, pH 7.4 / 26 W CFL bulb, r.t. → Biomolecule—NH_2

Scheme 1.9. Highly biocompatible reduction of an azide to an amine group.

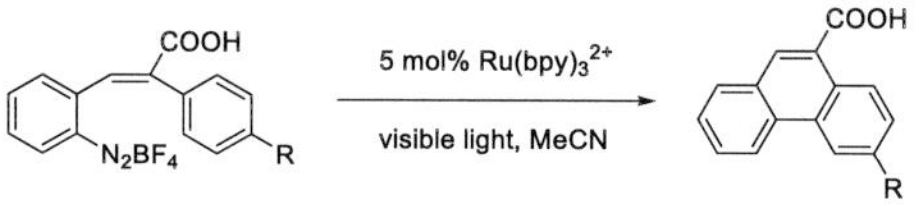

22 → 5 mol% $Ru(bpy)_3^{2+}$ / visible light, MeCN → **23**

Scheme 1.10. Photocatalytic Pschorr reaction.

Additionally, not only were existing reaction protocols improved, but also new synthetic transformations to construct bonds unthinkable by thermal reaction became possible.[45] One example for this is the ruthenium-catalyzed light-driven intramolecular radical anion [2+2] cycloaddition reaction (Scheme 1.11). While [2+2] cycloadditions do not necessarily require a catalyst, the reaction rate and selectivity can be improved significantly using photoredox catalysis.

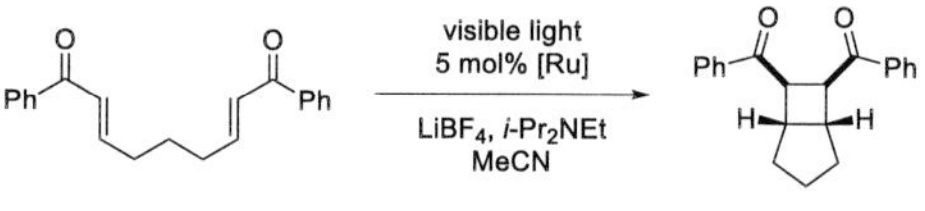

24 → visible light, 5 mol% [Ru] / $LiBF_4$, *i*-Pr_2NEt, MeCN → **25**

Scheme 1.11. Intramolecular radical anion [2+2] cycloaddition reaction.

1.2.2 Mechanisms of Photoredox Catalysis

The simple addition of photons by shining light on a suitable reaction setup starts a complex cascade of energy and electron transfer processes that are described in more detail in this section. A typical setup consists of photocatalyst, substrate, and sacrificial electron donor/acceptors if necessary. The most prominent photoactive molecule used is $Ru(bpy)_3^{2+}$, which shall serve here as an exemplary molecule, but also iridium, copper and metal-free photocatalysts are known.[46]

1.2.2.1 Photoexcitation

Visible light does usually not interfere with chemical reactions, as most bonds absorb only in the UV part of the spectrum. For this reason, a photoactive substance, also called photocatalyst, needs to be present that absorbs light in the visible range rendering it a colored compound. This photocatalyst is excited and becomes a more reactive intermediate, ready to transfer energy (then called photosensitizer) or an electron to the substrate.[46] The excitation proceeds by absorption of a photon and subsequent promotion of an electron from the ground state S_0 to one of the excited vibronic states of the singlet state S_1 (Figure 1.7). Relaxation of this excited state proceeds through non-radiative relaxation processes. From the S_1 ground state relaxation can occur either through fluorescence, the emission of a photon to relax back to S_0 or through intersystem crossing, the conversion of the singlet state to a vibronic excited triplet state T_1. From this triplet state, a rather slow because spin-forbidden emissive relaxation can occur, which is called phosphorescence.[47]

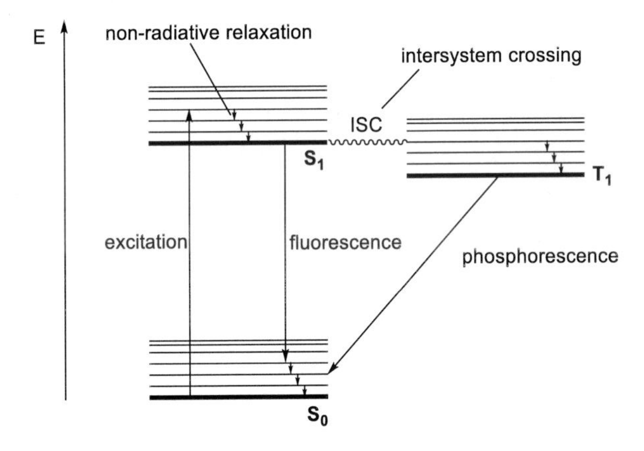

Figure 1.7. Jablonski diagram.

$Ru(bpy)_3^{2+}$ absorbs visible light with an absorption maximum at λ = 452 nm, which corresponds to a metal-to-ligand charge transfer (MLCT, Scheme 1.12). This corresponds to the excitation including intersystem crossing to form the long-lived triplet excitation state T_1. More visually, this can be thought of as an electron from the ruthenium center being transferred onto one of the bipyridine ligands. This charge separation makes the excited $[Ru(bpy)_3^{2+}]^*$ simultaneously a stronger oxidant and a stronger reductant than the ground-state. Thus, it can be either an energy donor, electron acceptor or electron donor, depending on the reaction it is used in.[46]

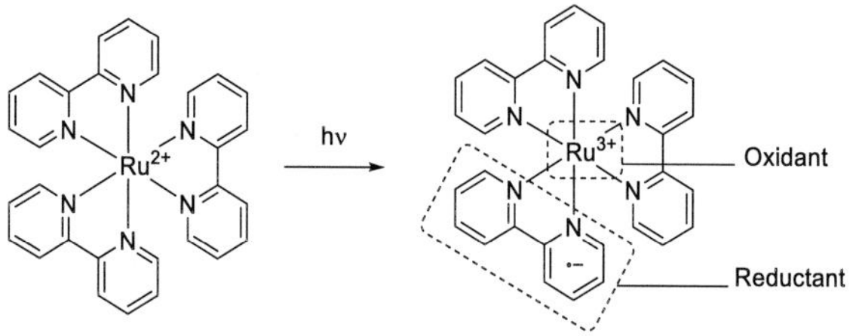

Scheme 1.12. Metal-to-ligand charge transfer upon photoexcitation of $Ru(bpy)_3^{2+}$

1.2.2.2 Quenching and Electron Transfer

After excitation, four scenarios are possible: (i) if a sacrificial electron acceptor Q is present, the catalyst is oxidized in an oxidative quenching cycle (Scheme 1.13, left). The generated photocatalyst cation PC^+ then acts as an electron acceptor itself to oxidize the substrate R in a single electron transfer process (SET) to complete the catalytic cycle. (ii) If however a sacrificial electron donor is present, the catalyst is reduced and in turn can reduce the substrate R in a reductive quenching cycle. (iii) If the substrate itself can be reduced by SET, a sacrificial donor completes the cycle (Scheme 1.13, right) or (iv) in case of a reductive quenching cycle, a sacrificial electron acceptor is needed to return the PC to its neutral ground state.

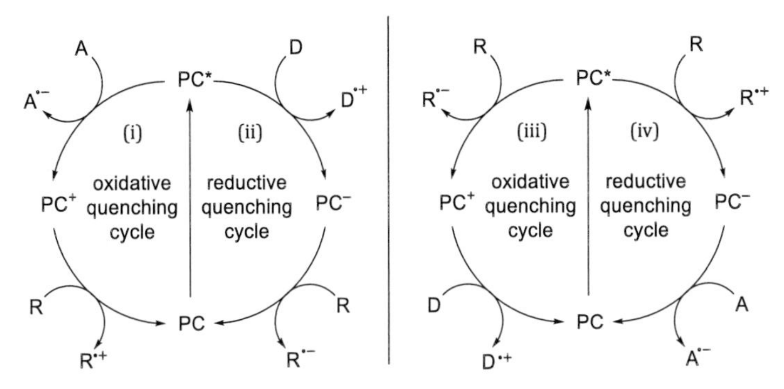

Scheme 1.13. Reductive and oxidative quenching cycles in photoredox catalysis.

If the reaction is self-sufficient so that no sacrificial SET reagents are needed, it is called redox-neutral.[45]

The most common sacrificial electron donors are aliphatic amines such as triethanolamine, diisopropylethylamine, or triethylamine.[48] Other common donors include the water-soluble ascorbic acid or easily oxidized thiols.[49] The typical electron acceptor is molecular oxygen.[48]

1.2.2.3 Bond formation

After the reactive intermediate has been formed by photoexcitation and the respective quenching mechanism, a myriad of reaction pathways is open that can afford the desired products by simple C–C bond or C–X bond formation, but also decarboxylative coupling, atom transfer radical addition, fluorination or many other mechanisms.[46]

1.2.3 Dual Photoredox Catalysis

Photoredox catalysis enables the use of reactive intermediates that are not accessible by other means like radical ions, diradicals or excited-state organic compounds. This becomes even more powerful when combined with traditional catalysis. The synergy between both strategies is achieved by activation/excitation by the photocatalyst and manipulation of that reactive intermediate by the second catalyst. While a lot of attention has been paid on the merger of photoredox catalysis with all kinds of traditional catalysis methods,[50] gold catalysis has especially seen a recent surge of interest.[51–54]

1.2.3.1 Photoinduced Electron Transfer

As explained in detail above, photoinduced charge separation leads to a state in which the transfer of electrons to or from a substrate occurs more easily than from the ground state. However, if the redox potential of the photocatalyst is not sufficient for a reaction with the substrate directly, a second catalyst can be used to mediate between photocatalyst and substrate, a so-called redox mediator (Scheme 1.14). Strategies to overcome this barrier include manipulating electrophilicity and reduction potential of heteroatom containing substrates by Brønsted/Lewis acid catalysis, organo- and transition metal catalysis.

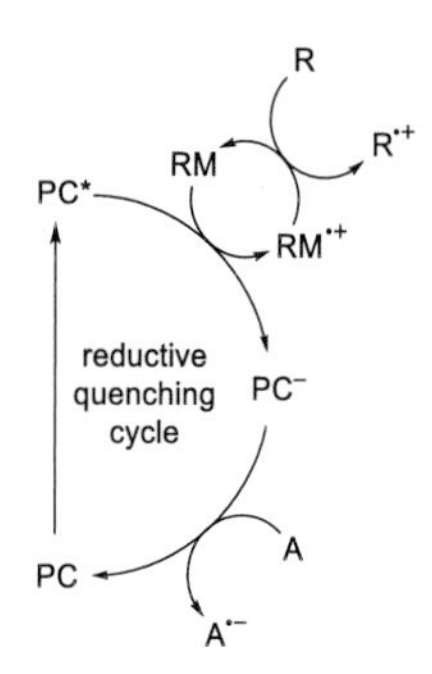

Scheme 1.14. Redox mediator in exemplary reductive quenching cycle.

1.2.3.2 Photoinduced Hydrogen Atom Transfer

Whenever a single electron transfer process happens, a radical or open-shell intermediate is generated. These open-shell intermediates need to be converted to closed-shell products to suppress unwanted side reactions. This can be done by hydrogen atom transfer (HAT) which is often achieved with an organic cocatalyst. Nicewicz and coworkers demonstrated this concept by an *anti*-Markovnikov alkene functionalization reaction (Scheme 1.15).[55] The generated product radical is not sufficiently oxidizing to propagate the radical chain, leading to very slow reaction rates. However, with the cocatalyst acting as a hydrogen atom shuttle, the reaction proceeds with dramatically increased rates.

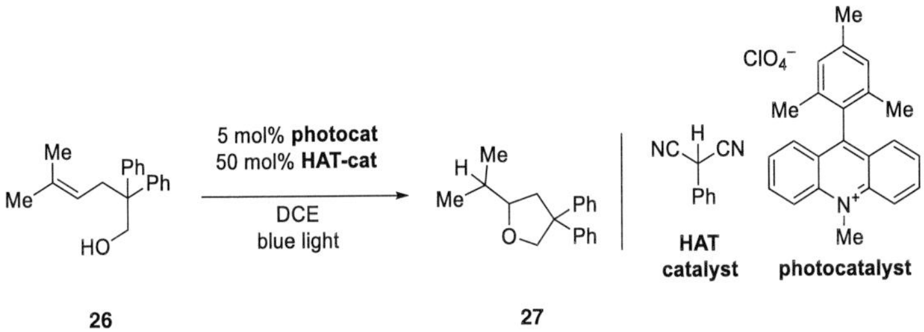

Scheme 1.15. Hydrogen atom transfer catalysis.

1.2.3.3 Photoinduced Energy Transfer

A third mechanism that can be co-catalyzed is the energy transfer from the excited state of a photocatalyst to the substrate by a mediating catalyst. This process is dominated by the Dexter energy transfer mechanism (Scheme 1.16). In the Dexter energy transfer, exchange of two electrons between two molecules leads to a transfer of energy without generation of charged species in the process. Several reports for dual catalysis energy transfer mechanisms exist, with most of them using the photocatalyst as a means of generating reactive singlet oxygen species.[56]

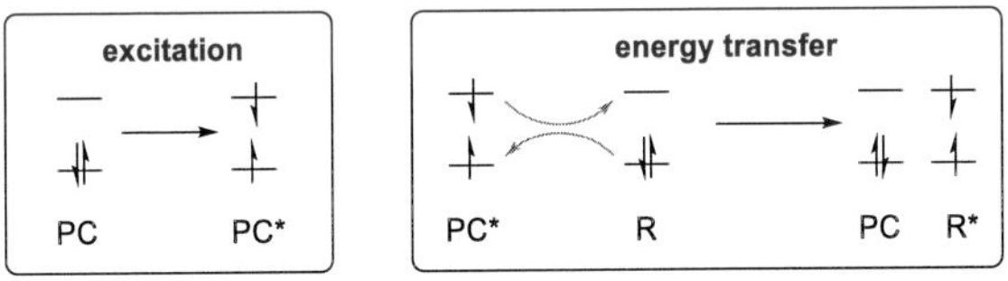

Scheme 1.16. Dexter energy transfer mechanism.

One exemplary dual catalysis employing singlet oxygen intermediates is the tandem allylic oxidation and epoxidation of an alkene (Scheme 1.17).[57] This reaction is catalyzed by a light-driven singlet oxygen generation from a porphyrin catalyst in conjunction with a titanium(IV) catalyzed epoxidation.

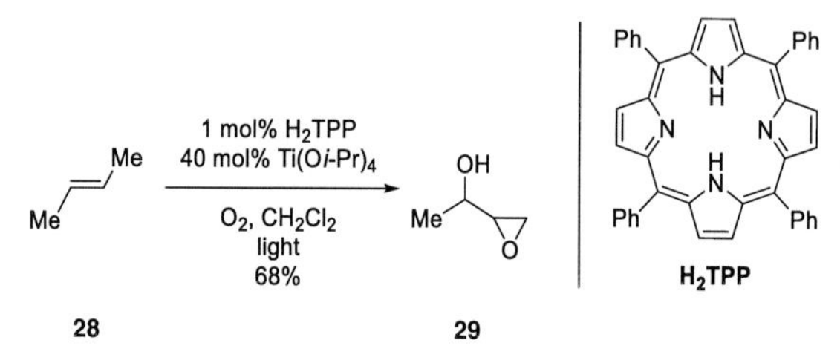

Scheme 1.17. Singlet oxygen generation *in situ* for the oxidation and epoxidation of alkene **28**.

2 AIM OF THIS WORK

In recent years, dual metal catalysis has seen a surge of interest. Specifically, the combination of one redox-active metal and a second, photon-capturing metal attracted great attention. The use of two metals working cooperatively offers more than the metals could achieve on their own. For instance, better atom economy, faster reaction rates, higher selectivity, and milder conditions can be achieved. Even new synthetic pathways that seemed impossible before, become accessible. However, the actual mode of cooperativity often remains quite nebulous. In particular, one important parameter that is poorly understood is the influence of the spatial relationship of the catalytic metal centers to each other. To overcome this, a defined, rigid, tunable and, first and foremost, synthetically accessible platform is needed. This would allow studies to be conducted about these effects.

[2.2]Paracyclophane (PCP), a rigid hydrocarbon, offers itself as such a platform. Being described as a "super-atom" with up to 16 precisely defined and fixated substituents, PCP's regioisomers allow for a tuning of geometrical relationships between two substituents (Figure 2.1).

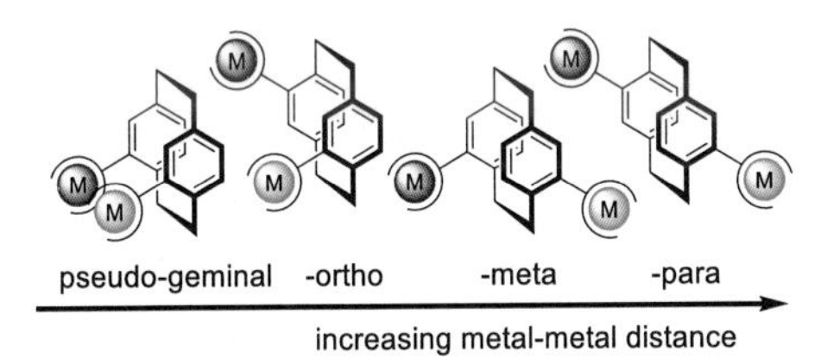

Figure 2.1. Distance modulation of two catalytic metal centers.

In this work, PCP is investigated for its suitability as a platform for distance-variable heterobimetallic complexes. In particular, an initial strategy on how to systematically prepare PCP heterobimetallic complexes is described. Furthermore, since PCP displays challenges in its functionalization, work is put into coming up with a modular solution for this. Next, *N*-donor ligands like 2-oxazolines, phenylpyridine and porphyrins are investigated as binding sites, because of their widespread use in coordination chemistry.

Then, after a general synthetic pathway to the desired heterobimetallic complexes has been established, a target complex will be designed and two isomers with a different metal-metal distance will be prepared. Here, both metal atoms are in a suitable coordination environment to serve as catalytically active centers in dual photoredox catalysis, optimally in a cooperative fashion. Finally, the probing of these model complexes in photoredox catalysis will serve to evaluate the viability of PCP as a backbone for distance-variable heterobimetallic complexes and their application in cooperative catalysis.

In conclusion, in this work PCP will be demonstrated to be a suitable platform to build defined, rigid and tunable heterobimetallic complexes (Figure 2.2). This platform can be used to study cooperative effects in catalysis. A photoredox catalysis reaction will serve as an example for this and emphasize the viability of this approach.

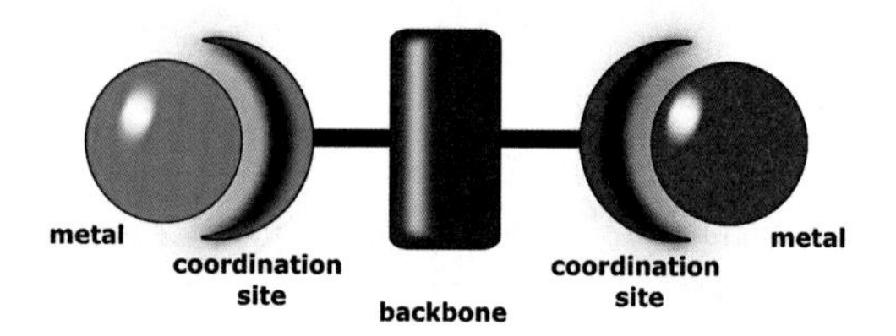

Figure 2.2. General structure of heterobimetallic complexes with a PCP backbone.

3 MAIN PART

3.1 Strategy

In this chapter, the aim of this work is explained in more detail and an overall strategy to access the heterodisubstituted PCPs required for the preparation of heterobimetallic complexes is presented.

3.1.1 Roadmap towards Heterobimetallic Complexes

3.1.1.1 Definition

Any well-defined metal complex that bears exactly two different metal atoms is classified as a heterobimetallic complex.[58] The general structure (Figure 3.1) comprises two different metal atoms, two different coordination sites and a linking backbone that defines the spatial relationship between the coordination sites and thus the metals.

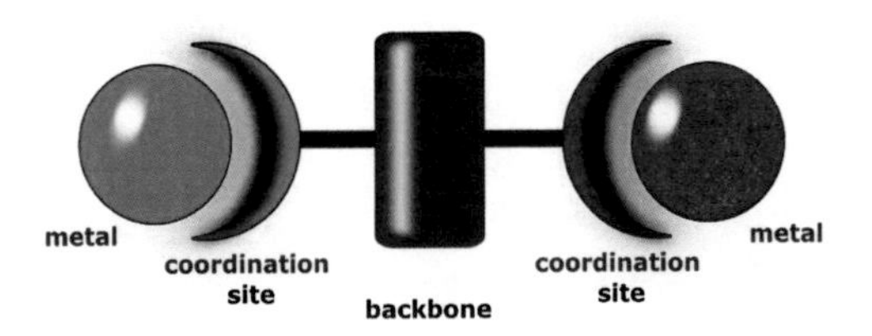

Figure 3.1. General structure of a heterobimetallic complex.

In the following sections, the structure is analyzed in detail and a rationale for the choices of backbone, coordination sites and metals in this work is presented.

3.1.1.2 The backbone

Heterobimetallic complexes are held together by a linking backbone that can range from a single bond between the coordination sites, up to the most sophisticated arrangements of structures, providing various functionalities such as structural rigidity, light-harvesting ability or electronic communication/isolating properties.

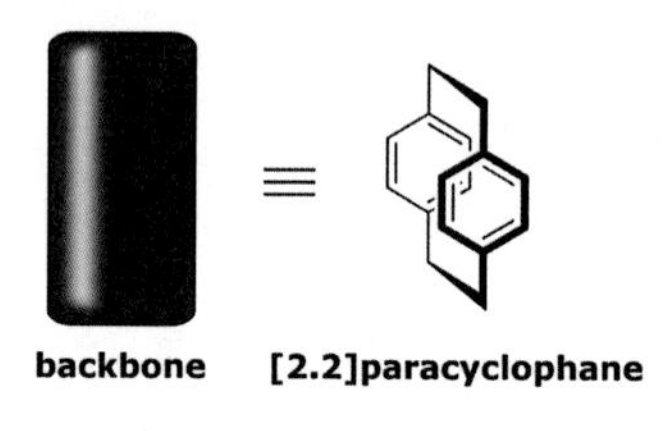

Figure 3.2. Choice of the backbone in this work: [2.2]paracyclophane.

This work uses PCP as a backbone and platform on which heterobimetallic complexes are constructed in a multistep synthesis (Figure 3.2). PCPs feature four definitive advantages for this plan: (i) The overlapping π-orbitals of the cofacial-stacked decks allow for a strong through-space electronic communication[5] and thus, possibly electronic communication between two metal atoms in the heterobimetallic complex. (ii) The inherent structural rigidity of PCP leads to a fixed spatial geometry of the substituents/coordination sites, which can be used to design distance-variable heterobimetallic complexes with a defined distance between the metal atoms. As the PCP structure allows for through-space conjugation,[6,59–61] it was decided to have one substituent (coordination site) per deck (Figure 3.7), to study this conjugation mode in more detail. All the available substitution patterns fulfilling this, namely pseudo-*para*, -*meta*, -*ortho* and -*geminal* allow to set the metal–metal distances in a stepwise fashion (Figure 3.3). (iii) Similar to ferrocene, substituted PCP exhibits planar chirality which is used as a core structure in asymmetric catalysis.[25] (iv) Using the large body of synthetic knowledge available for the synthesis of PCP derivatives, a large library of ligand motifs is accessible.

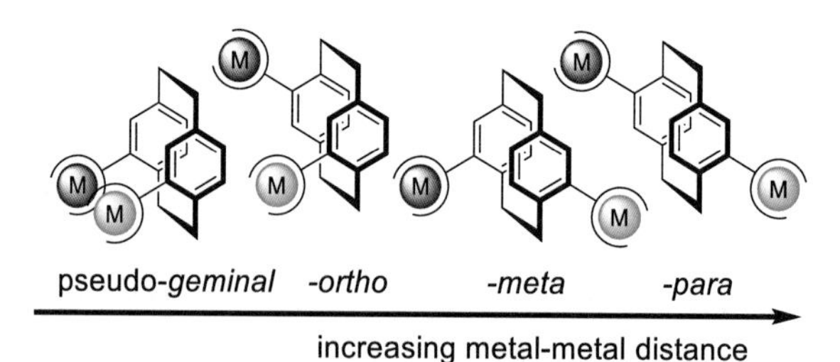

Figure 3.3. Stepwise increase of the metal-metal distance *via* a change in substitution pattern of [2.2]paracyclophane.

3.1.1.3 Coordination Site and Metal Center

If the backbone from the previous section is equipped with two different binding sites, the arrangement will become a heteroditopic (=comprising two different binding sites) ligand. This structure is now capable of bridging two different metal atoms by forming one or multiple bonds between coordination site and the free metal atom (Figure 3.4).

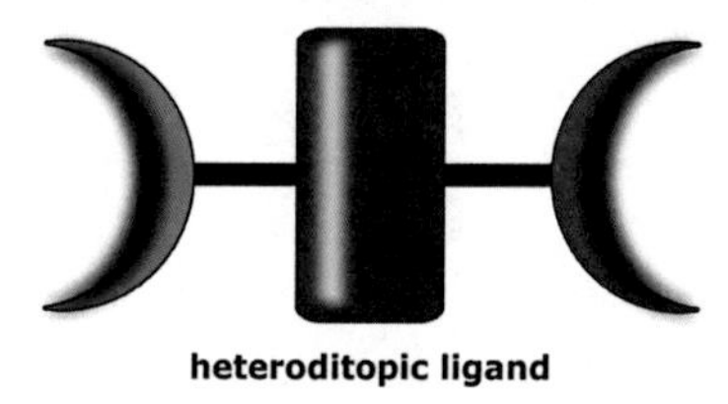

Figure 3.4. The backbone including two different binding sites becomes a heteroditopic ligand and can bind two different metal atoms.

The exact design of each coordination site comes with the following requirements (Figure 3.5): (i) A synthetic access must exist to construct a coordination site on a given backbone. Even if synthetic access is possible, synthetic simplicity aids dramatically in realizing the coordination site motif. (ii) The introduction of the second coordination site must be conducted in conditions that are compatible with the first coordination site. (iii) For the construction of a well-defined heterobimetallic complex, the coordinating atoms must bind selectively to one of the metal atoms that are introduced such that site-selective coordination occurs. (iv) Strong binding between coordination site and metal is required to avoid decomplexation or the formation of mixed species including homobimetallic complexes (metal–metal scrambling).[62]

(i) ease of synthetic access

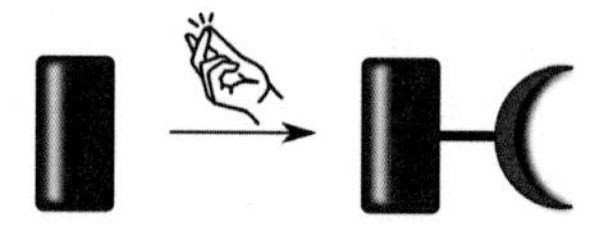

(ii) tolerance of second functionalization

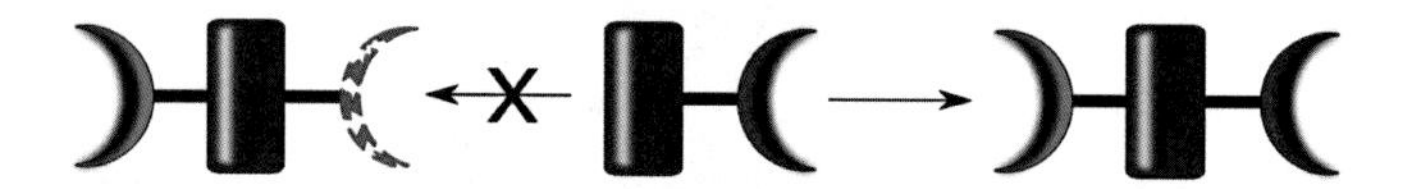

(iii) metal - coordination site match

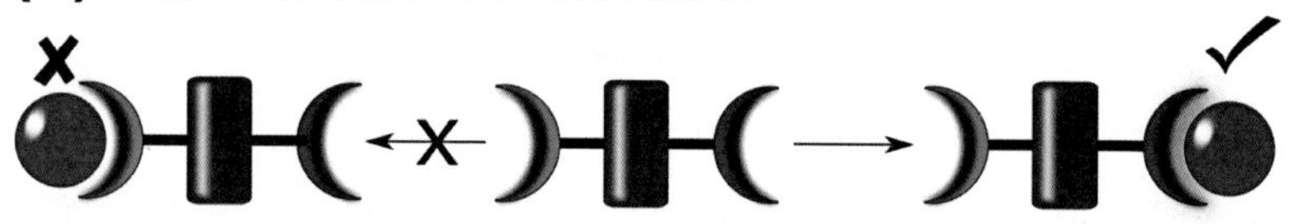

(iv) complexation stability

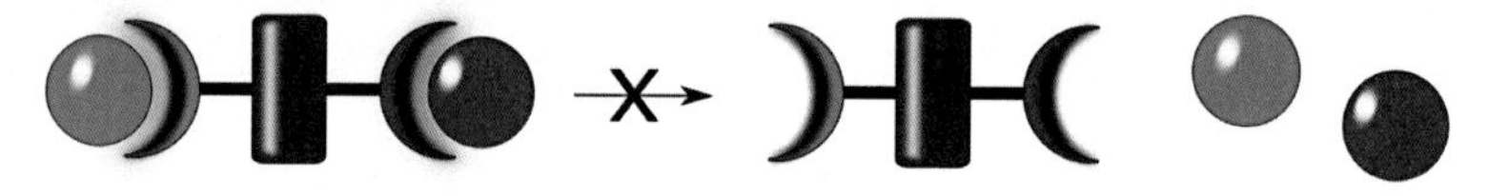

Figure 3.5. Requirements that must be met for the successful coordination site design to synthesize heterobimetallic complexes.

3.1.1.4 Coordination Site and Metal Choice

For this work, two ligand motifs were chosen for closer investigation: P-donor triaryl phosphines and *C,N*-chelating aryl pyridines.

Phosphines are more readily available than *N*-heterocyclic carbenes (NHCs), a comparable class of spectator ligands, fulfilling requirement (i) from the previous section 3.1.1.3. Furthermore, NHCs were discussed as phosphine analogues for some time and are now regarded as a separate ligand class, due to their complex chemistry and inability to "correct binding errors" after initial coordination.[63] While phosphines themselves are prone to oxidation, they can readily be converted to phosphine oxides[64] which are stable enough for the introduction of another coordination site by organolithium reagents or cross-coupling methodologies, fulfilling requirement (ii). Concerning requirement (iii), phosphines are known to form complexes with all transition metals[65] and thus the phosphine has to be metallated first to protect it from the accidental and undesired coordination to the second metal atom. Then, the second coordination site can be metallated to ensure two different metals are inserted. However, the soft character of phosphines and gold(I) ions make them a perfect match for stable (requirement iv) coordination compounds. Phosphines show additional π-acidity allowing π-backbonding, thus increasing the P–M bond strength. It should be noted though that in recent years the exact nature and strength of π-backbonding has been subject to discussion.[66] Phosphines can be coordinated to gold atoms to generate very stable gold phosphines. Especially gold(I) has seen increasing attention for application in general catalysis,[51,67,68] visible-light mediated photoredox catalysis[52,69–72] and anticancer applications.[73]

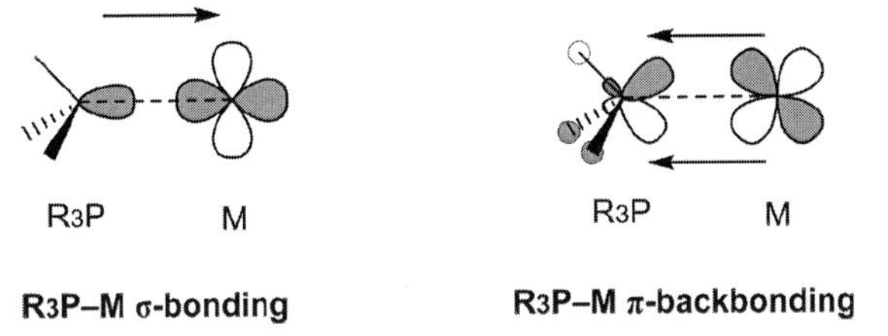

Figure 3.6. Orbital interactions in metal-phosphine bonding and backbonding.[65]

Chelating *C,N* coordination sites like phenyloxazolines[74] or phenylpyridines[30,32] that undergo cyclometallation are an excellent choice for a complementation to the phosphine coordination site. Both of these ligand motifs have been reported to form cyclometallated complexes with ruthenium[24,34,75,76] and palladium,[77–79] which are interesting metal centers for applications in catalysis.

3.2 Differentiation and Synthesis of Heterodisubstituted PCPs

The overarching goal of synthesizing heterobimetallic complexes was to be achieved by preparation of a heteroditopic (i.e. containing two different binding sites) ligand that uses the PCP as a linking backbone. These compounds are inherently asymmetric (heterodisubstituted) with respect to the substitution pattern at the PCP. Given the different synthetic paths to the possible substitution patterns, this heterodisubstitution must be targeted on a pattern-individual basis.

The pseudo-*geminal* substitution pattern is the only one that is synthesized from the prefunctionalized monosubstituted PCP, whereas all other geometries are accessed by the respective dibromide (Scheme 3.1).

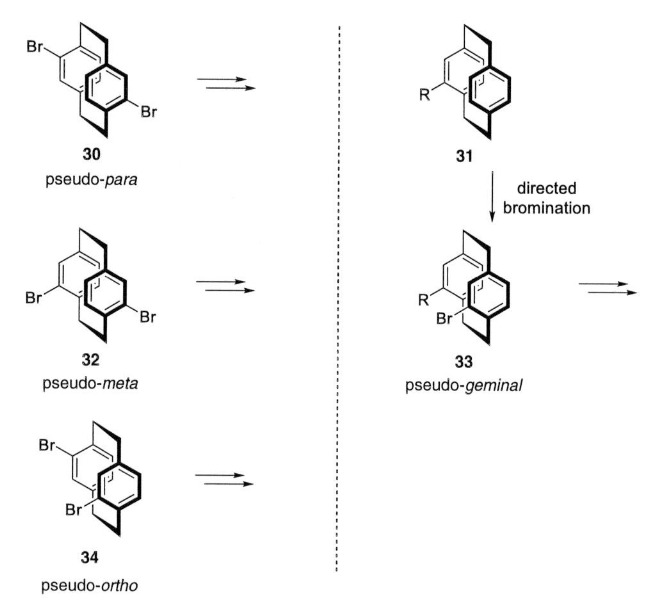

Scheme 3.1 Access to the pseudo-substitution patterns starting from the respective dibromides **30**, **32** and **34**. Pseudo-*geminal* **33** is accessible by bromination of prefunctionalized PCP **31**.

This poses the challenge of going from a homodisubstituted compound (Scheme 3.1, **30**, **32**, **34**) to a heterodisubstituted one (Figure 3.7). As the two functional groups in the dibromides are identical, the only way to address this issue is by means of adjusting stoichiometry and selectivity in order to transform only one bromide into a different functional group. In sections 3.2.1.1 and 3.2.1.2 methods to achieve this differentiation are discussed.

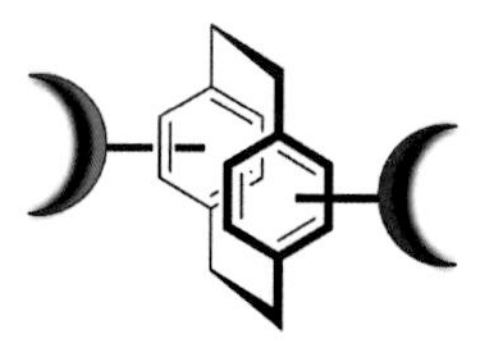

Figure 3.7 Heteroditopic ligand with PCP as a linking backbone. An asymmetric structure that bears two different binding sites, one on each deck.

3.2.1.1 Cross-Coupling

One of the most common transformations for aryl bromides are cross-coupling reactions. In these metal-catalyzed reactions, the aryl bromide bond is broken and in turn a new aryl–carbon or aryl–heteroatom bond is formed, thus extending the carbon skeleton.

An example from literature for partial cross-coupling of aryl dibromides points towards difficulty in getting high yields for the heterodisubstituted product **37** (Scheme 3.2).[80] The same is true for PCP dibromide **30** in a cross-coupling reaction with stannane **39**, resulting in significant amounts of homodisubstituted side-product **41** instead of desired **40**.

Br
Br
+
$B(OH)_2$
Suzuki
cross-coupling
Br
+
35 **36** **37** (22%) **38** (61%)

Scheme 3.2 Cross-coupling of 1,4-dibromobenzene leads to more homodisubstituted product **38** than heterodisubstituted product **37**.

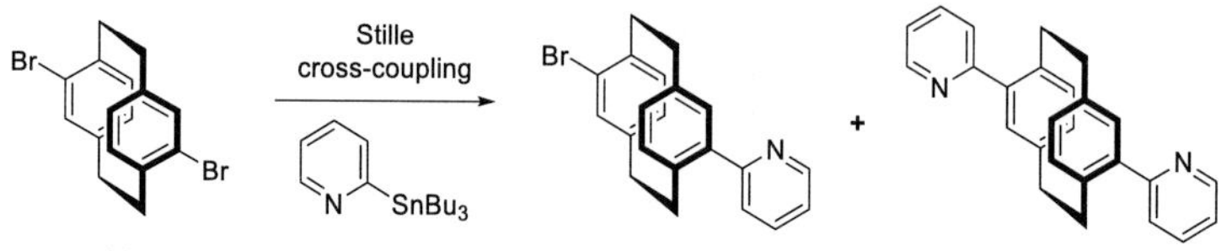

Stille cross-coupling

30 39 **40** (43%) **41** (27%)

Scheme 3.3 Stille cross-coupling reaction of the dibromide **30** with stannane **39** delivers only moderate yields for the heterodisubstituted product **40**.

Alternatively, if the heterodisubstitution pattern is set up before the cross-coupling step, as shown on the generic PCP **42** (Scheme 3.4), the problem of homocoupling can be avoided. Thus, a method to prepare compounds such as **42** is very valuable.

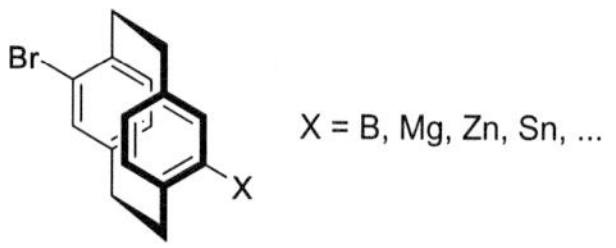

X = B, Mg, Zn, Sn, ...

42

Scheme 3.4 Compounds **42** with a heterodisubstituted pattern, suitable for selective cross-coupling.

3.2.1.2 Selective Monolithiation

An alternative to cross-coupling reactions for the formation of new C–C bonds from aryl halides is presented in this section. Aryl halides can undergo a metal-halogen exchange with organolithium reagents. The driving force of this reaction is the formation of a more stable organolithium compound. To this end, reactive alkyllithium reagents are used to transform aryl halides into relatively stable aryllithium reagents with alkyl halides as byproduct. These newly formed aryl lithium reagents are usually not stable enough for isolation, thus demanding trapping of the organolithium with a suitable electrophile. Therefore, an aryl–C or aryl–heteroatom bond is formed, and the now stable product can be isolated and purified by standard organic procedures.

However, upon lithiation of PCP dibromide **30**, a statistical distribution is expected, comprised of the desired heterodisubstituted 4-bromo-16-lithio[2.2]paracyclophane (**43**, 50%), twice reacted 4,16-dilithio[2.2]paracyclophane (25%) and remaining 4,16-dibromo[2.2]paracyclophane (**30**, 25%) is obtained (Figure 3.8). However, experimental results showed that nearly quantitative conversion to the desired heterodisubstituted product is achieved.

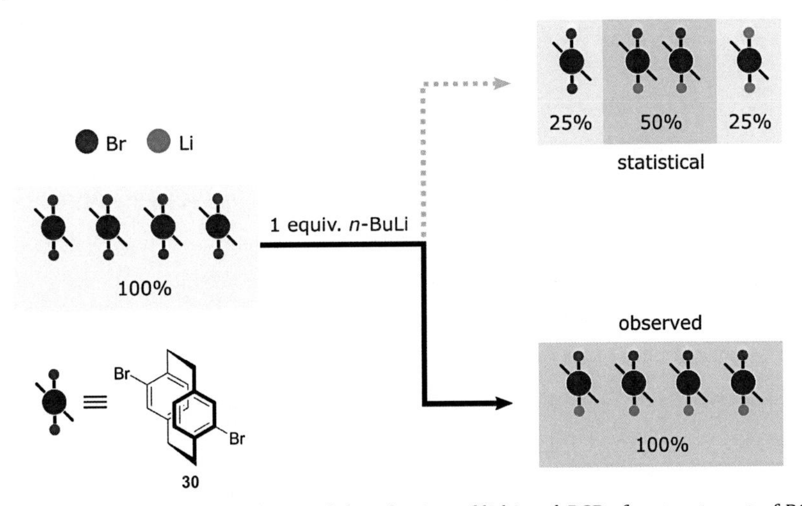

Figure 3.8. Statistical vs. observed distribution of lithiated PCP after treatment of PCP dibromide **30** with 1 equiv. of *n*-BuLi.

The metal-halogen exchange on an already lithiated substrate must be significantly less favored compared to the initial metal-halogen exchange at the PCP dibromide, hence resulting in the observed distribution. Hints toward this behavior have been published previously[11,59,81] but no in-depth study has been conducted. Part of this work is reported in the master's thesis preceding this work.[64] It was found that a reaction time of 30 minutes in THF at –78 °C and the use of 1.20 equivalents of *n*-BuLi delivered the best yields after electrophilic trapping of the monolithio-intermediate. Longer reaction times slowly increased the amount of twice reacted 4,16-dilithio[2.2]paracyclophane, while shorter reaction times led to incomplete conversion of starting material **30**.

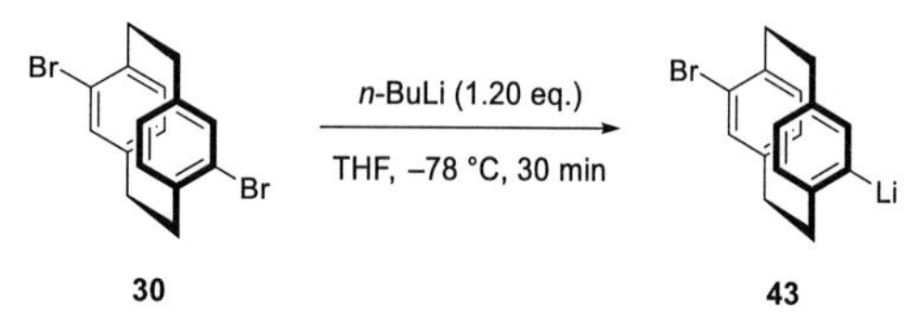

Scheme 3.5 Optimized conditions for monolithiation of PCP dibromide **30**.

Using this protocol, a range of heterodisubstituted PCPs are accessible as summarized in Table 1. The excellent yields for phosphine oxide **44** or trifluoroborate **49** is good evidence for the significantly higher presence of the heterodisubstituted intermediate **43** than the statistically expected 50% (Figure 3.8). The aldehydes accessible by this protocol (Table 1, entry 2 and 3) are an excellent entry point to enantiomerically pure PCPs and have a rich follow-up chemistry, as is demonstrated in section 3.5.2. The trifluoroborates (Table 1, entry 6 and 7) on the other hand are excellent cross-coupling substrates and are discussed in detail in section 3.3.

Table 1. Heterodisubstituted PCPs accessible by selective monolithiation and subsequent electrophilic trapping.

Br, Br → Br, E
1) *n*-BuLi (1.20 equiv.) –78 °C, THF, 30 min
2) electrophile "E" –78 °C to r.t., overnight
30,34 → **44-50**

Entry	Electrophile	Product	Label	Yield[a] [%]
1	PPh_2Cl	Br, $P(O)Ph_2$	**44**	96
2	DMF	Br, CHO	**45**	75
3	DMF	Br, OHC	**46**	55

4[64]	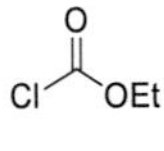 Cl–C(O)–OEt	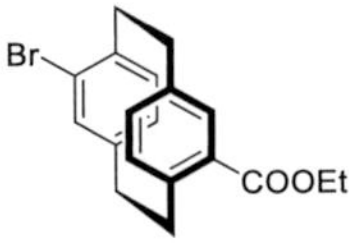 Br, COOEt	**47**	60
5[64]	CO_2	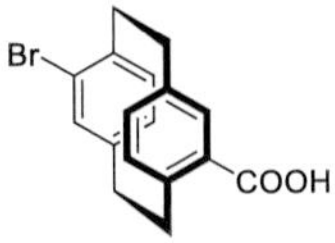 Br, COOH	**48**	traces
6	$B(OMe)_3$, then KHF_2	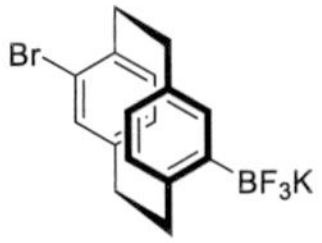 Br, BF_3K	**49**	91
7	$B(OMe)_3$, then KHF_2	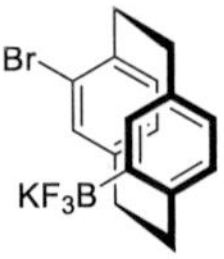 Br, KF_3B	**50**	66

[a] isolated yields.

3.2.2 Conclusions

This section dealt with the underlying issue of synthesizing heterobimetallic complexes. A synthetic target has been set to aim for phosphine coordination sites that would bind to gold(I) combined with C–N ligands suited for cyclometallation with palladium and ruthenium. The synthetic access starts with the backbone dibromoPCP, which is converted to a heterodisubstituted PCP by a newly developed selective monolithiation protocol. In the further sections of this work, PCP is further functionalized to generate free binding sites by a range of cross-coupling (with new protocols developed in this work) and organolithium/electrophile trapping transformations. Ultimately, the metal atoms are introduced in a stepwise fashion to ensure the precise construction of heterobimetallic complexes.

3.3 PCP Trifluoroborates in Suzuki Cross-Coupling

The Suzuki-Miyaura or simply Suzuki cross-coupling reaction is arguably the most modern method to forge new carbon–carbon bonds.[82] Among the most challenging substrates for the Suzuki reaction are bulky *ortho*-substituted or *ortho,ortho'*-substituted aryls.[83] Since PCP bears substituents in the *para* positions of both benzene rings, any additional carbon–carbon bond formation on the benzene rings has to deal with the difficulties of a bulky *ortho*-substituted cross-coupling reaction. Additionally, PCP is renowned for its often sluggish reaction behavior,[1] which is especially pronounced in reactions that depend significantly on the electronic and steric properties of the reaction center. The steric hindrance by the ethylene bridges and its less-than-aromatic character discussed in section 1.1 render cross-couplings involving PCP cumbersome.

While Suzuki cross-coupling reactions with PCP are known, they focus almost exclusively on PCP halides and respective boron cross-coupling partners. This is however a subpar setup, as PCP halides are very electron-rich compounds, albeit the rate limiting oxidative addition step is facilitated by electron-deficient organohalides.[84] Thus an inversion with the PCP bearing the boron moiety should lead to far superior results.

In this chapter, a short overview over available cross-coupling protocols for PCP is given, followed by a more in-depth look at Suzuki cross-coupling reactions. A new class of PCP Suzuki cross-coupling substrates is presented that overcome a range of issues (regarding preparation, storage, handling, toxicity, versatility, and yield) of previously reported PCP cross-coupling substrates: PCP trifluoroborates.

3.3.1 Cross-Coupling of PCP

In this section an overview of the literature known protocols for palladium-catalyzed cross-coupling involving the PCP moiety is given.

3.3.1.1 General Cross-Coupling Mechanism

To understand the more specialized cross-coupling reactions, the general mechanism needs to be discussed. The catalytic cycle starts with an active catalyst in resting state, often in oxidation state 0 (Scheme 3.6). In the first step, oxidative addition of the aryl halide affords the metal-inserted intermediate that is ready to undergo transmetalation with an organometallic compound M'-R'.

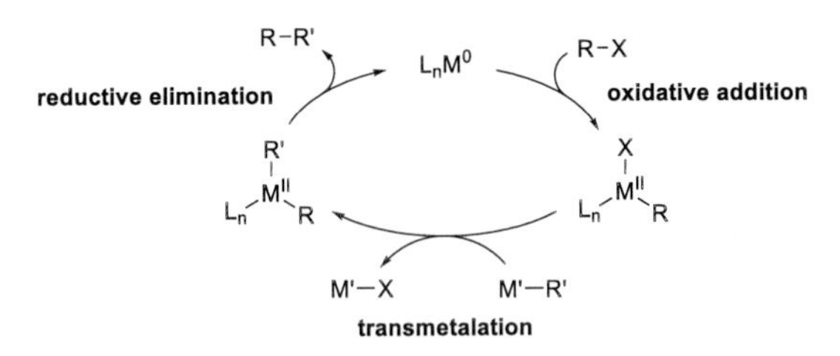

Scheme 3.6 General cross-coupling mechanism.

The resulting intermediate reductively eliminates to afford the cross-coupling product and returns the catalyst to its resting state. The exact makeup of organometallic species and cross-coupling aryl halide partner determine the name of the cross-coupling reaction. Some common variants are discussed in the following sections.

3.3.1.2 PCP Kumada cross-coupling

The Kumada cross-coupling reaction describes the C–C bond formation between an organomagnesium reagent and an organohalide by means of metal-catalyzed cross-coupling.[85] While palladium catalysis is the most commonly described variant, nickel catalysis is known as well.[86] The reaction overview is shown in Scheme 3.7.

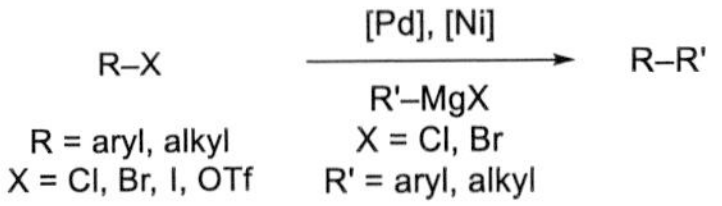

$$\text{R–X} \xrightarrow[\text{R'–MgX}]{\text{[Pd], [Ni]}} \text{R–R'}$$

R = aryl, alkyl
X = Cl, Br, I, OTf

X = Cl, Br
R' = aryl, alkyl

Scheme 3.7 Reaction overview for palladium or nickel-catalyzed Kumada cross-coupling reactions.

This methodology has been successfully applied to the arylation of PCP bromide **51** with various sterically crowded aryl magnesium bromides by Zemanek and Kus in 1985.[87]

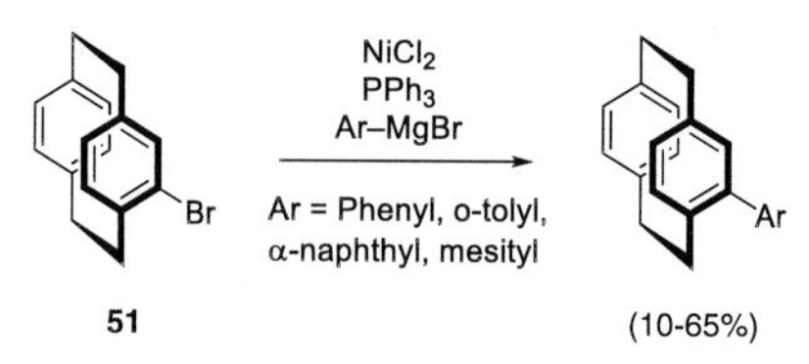

Scheme 3.8 Nickel-catalyzed Kumada arylation of PCP bromide **51**.

Recently, the Kumada cross-coupling was used for the heteroarylation of electron-rich para-methoxy iodo-PCP **52** with 2-pyridyl magnesium bromide (**53**) to form biaryl **54** (Scheme 3.9) albeit in not satisfying yields.[88] Thus, a better synthetic strategy to access valuable biaryl **54** is necessary.

$Pd_2(dba)_3$
$(t\text{-}Bu)_2PHO$
THF, 70 °C, 40 h
I
MeO
MgBr
N
52 **53** **54** (16%)

Scheme 3.9 Successful synthesis of heteroarylated *para*-disubstituted PCP **54**.

The major drawback of the Kumada cross-coupling is the reactivity of the organomagnesium species. Thus, substrates bearing functional groups not tolerating this highly reactive cross-coupling intermediate, such as aldehydes, ketones, reactive esters or nitriles, are not suitable for this methodology. Depending on the organic substituent of the organomagnesium species, storage can also be problematic as compounds of this type can have very short shelf life time and therefore must be prepared freshly before each use.

3.3.1.3 PCP Negishi cross-coupling reaction

The Negishi cross-coupling reaction describes the C–C bond formation between an organozinc reagent and an organohalide by means of metal-catalyzed cross-coupling. Similar to the Kumada cross-coupling, palladium catalysis dominates the field[89] but other metals, such as copper[90], cobalt[91,92] and nickel,[93-97] have been reported to catalyze this transformation as well (Scheme 3.10). This reaction type has been applied to the paracyclophane only in one case (same reaction as Scheme 3.9, R-**Zn**Br instead of

R-**Mg**Br) and with limited success, as the yields were comparable to the organomagnesium approach.

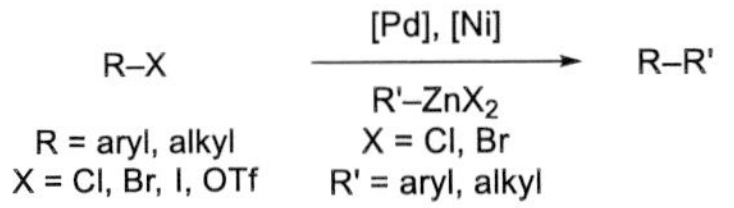

$$\mathrm{R{-}X} \xrightarrow[\mathrm{R'{-}ZnX_2}]{[\mathrm{Pd}],\ [\mathrm{Ni}]} \mathrm{R{-}R'}$$

R = aryl, alkyl
X = Cl, Br, I, OTf

X = Cl, Br
R' = aryl, alkyl

Scheme 3.10 Reaction overview for Negishi cross-coupling reactions.

Like the Kumada cross-coupling reaction, the drawbacks of this reaction are found again in the reactivity of the organozinc species which are less reactive than the organomagnesium reagents used in the Kumada cross-coupling but still not compatible with sensitive functional groups.

3.3.1.4 PCP Stille cross-coupling

The Stille cross-coupling reaction describes the C–C bond formation between an organotin reagent and an organohalide by means of metal-catalyzed cross-coupling.[98] The reaction is almost exclusively mediated by palladium, but additives of other metals like lithium or copper are very common.[99] Copper additives enhance the reaction rates by a factor of >10^3.[100] Two possible modes of enhancing the Stille reaction are discussed in literature. Copper can act as a phosphine scavenger, leading to faster rates by precluding "autoretardation" during the transmetalation step by free phosphines.[99] Additionally, transmetalation of the organotin species to generate an organocopper species has been discussed as well, with some examples see in a palladium replacement possible through the use of copper catalysis.[99] Lithium additives (most often lithium chloride) can stabilize the transition state of the oxidative addition step and improve the transmetalation step rate by increasing the solvents polarity, although depending on substrate, ligand system and solvent also retardation through lithium chloride has been reported.[100,101] The reaction overview is shown in Scheme 3.11.

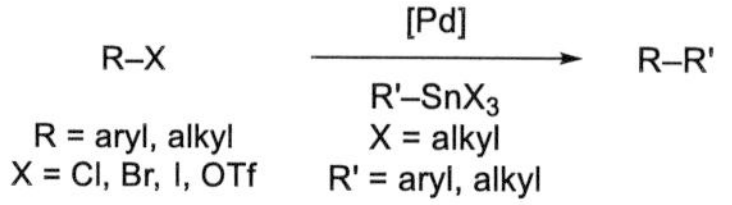

R–X → [Pd], $R'–SnX_3$ → R–R'

R = aryl, alkyl
X = Cl, Br, I, OTf

X = alkyl
R' = aryl, alkyl

Scheme 3.11 Reaction overview for Stille cross-coupling reactions.

Since organotin reagents are not as reactive as organomagnesium or -zinc derivatives, the Stille reaction is much more tolerant of functional groups.[99] This method was used successfully to arylate PCP both with PCP-organotin species and aryl halides as well as cross-coupling of PCP halides with organotin coupling partners.[88,102,103] Braun *et al.* could synthesize the 4-(2'-pyridyl)[2.2]paracyclophane **15**, a biaryl mimicking a widespread structural motif in coordination chemistry (i.e. phenylpyridine) with their Stille cross-coupling protocol (Scheme 3.12).[88]

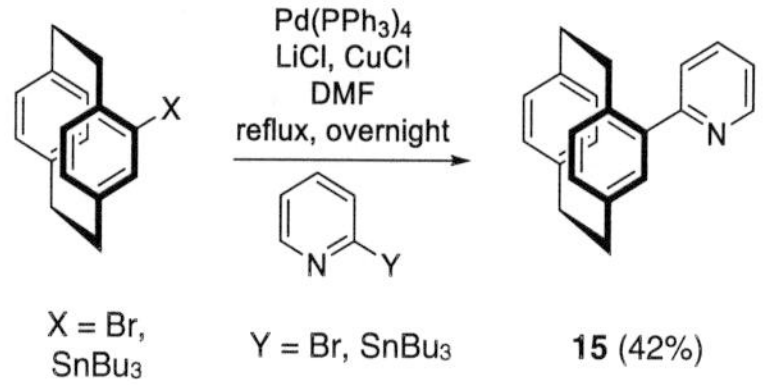

$Pd(PPh_3)_4$
LiCl, CuCl
DMF
reflux, overnight

X = Br, $SnBu_3$

Y = Br, $SnBu_3$

15 (42%)

Scheme 3.12 First Stille cross-coupling approach to heteroarylated PCP **15** mimicking the phenylpyridine motif.

3.3.1.5 PCP Suzuki cross-coupling

The Suzuki cross-coupling reaction describes the C–C bond formation between an organoboron reagent and an organohalide by means of metal-catalyzed cross-coupling assisted by a base.[82] Traditionally catalyzed by palladium, in recent years the field has seen a slow transition towards more sustainable nickel catalysis.[104] In addition, more efforts were made to encompass catalytic systems comprised of gold, iron, copper, silver and cobalt, to ultimately make this reaction suitable for green chemistry, preferably being run with highly efficient catalysts in green media.[105] The reaction overview is shown in Scheme 3.13. Organoboron derivatives share with organotin derivatives that they are very tolerant towards sensitive functional groups.[82]

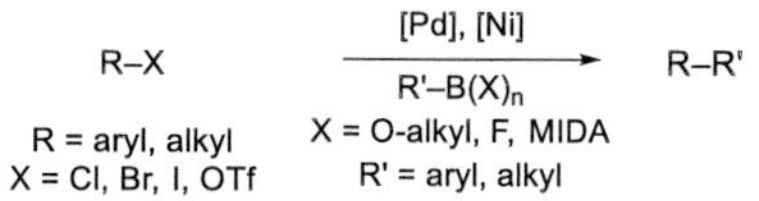

[Pd], [Ni]
R–X
R–R'
R'–B(X)n
R = aryl, alkyl
X = Cl, Br, I, OTf
X = O-alkyl, F, MIDA
R' = aryl, alkyl

Scheme 3.13 Reaction overview for Suzuki cross-coupling reactions.

In 2004 Roche *et al.* published a seminal report on Suzuki cross-coupling reactions involving PCP.[106] Successful reactions were performed with 4-bromo[2.2]paracyclophane (**51**) as the electrophilic cross-coupling partner (Scheme 3.14).

Br
Suzuki
cross-coupling
Ar-B(OR)2
R = H, Alkyl, pin
Ar

51

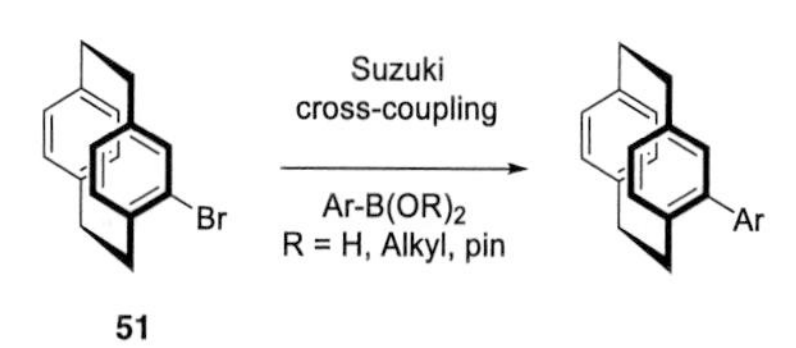

Scheme 3.14 Suzuki cross-coupling reactions with bromoPCP **51**.

However, upon reversing the roles of nucleophile and electrophile, i.e. using PCP boron derivatives, the following issues arose: PCP boronic acid is inherently unstable and decomposes to the corresponding alcohol. PCP boronic esters are unstable or in case of the pinacol ester lead to cross-coupling yields <10% after one week at elevated temperatures (Scheme 3.15).

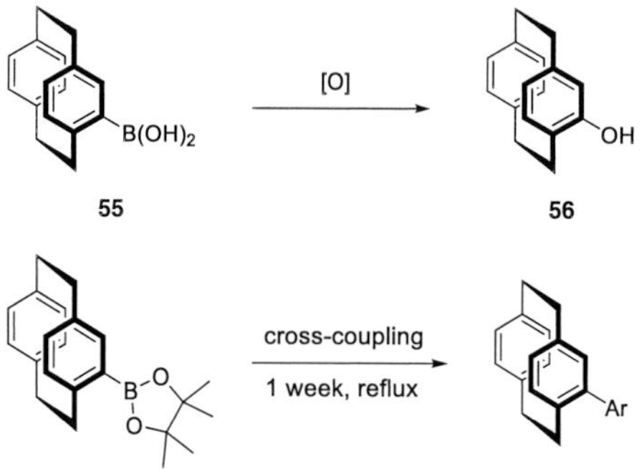

Scheme 3.15. Decomposition of boronic acid or boronic esters **55** to the corresponding alcohol and sluggish reactivity of pinacol ester **57**.

These issues of boronic acids are specific to PCP chemistry, but even in the more general sense, boronic acids are far from ideal reagents: Often they are difficult to purify solids of wax-like consistency and form trimers that are cyclic anhydrides of boronic acids (boroxines, Scheme 3.16), which imparts reactivity and requires superstoichiometric use of boronic acids.[107]

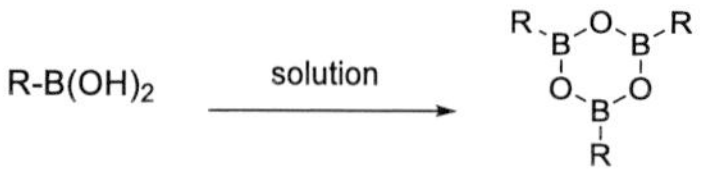

Scheme 3.16. Formation of boroxines **60** from organoboronic acids **59** in solution.

However, organotrifluoroborates were reported by Vedejs *et al.* as convenient precursor for boronic acids.[108] Their application as cross-coupling partners was investigated by Darses *et al.*[109] For a while no attention was paid to the then less reactive aryl bromides, tosylates and iodides. Only in 2003 this was investigated more in detail by Molander *et al.*, who published a palladium-catalyzed cross-coupling of organotrifluoroborates with aryl bromides.[110]

3.3.1.6 Mechanism of the Suzuki cross-coupling reaction with trifluoroborates

Suzuki cross-coupling reactions follow the general mechanism for palladium-catalyzed cross-coupling reactions as discussed above. The use of base is necessary in Suzuki cross-couplings as not the boronic acid itself but the borate anion (base-promoted reduction) is the active transmetalating agent.[111]In the case of a reaction where substrate A is converted to intermediate B with rate constant k_1 and then finally product C is formed from B with rate constant k_2 (Scheme 3.17), two possible cases arise: In one case, where $k_1 > k_2$, B is formed faster than it is consumed, leading to accumulation of intermediate B. In the other case, where $k_1 < k_2$, intermediate B is consumed nearly instantly after its formation to afford product C (Figure 3.9).

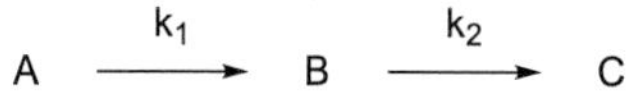

$$A \xrightarrow{k_1} B \xrightarrow{k_2} C$$

Scheme 3.17. Reaction of substrate A being converted to intermediate B forming product C with the rate constants k_1 and k_2.

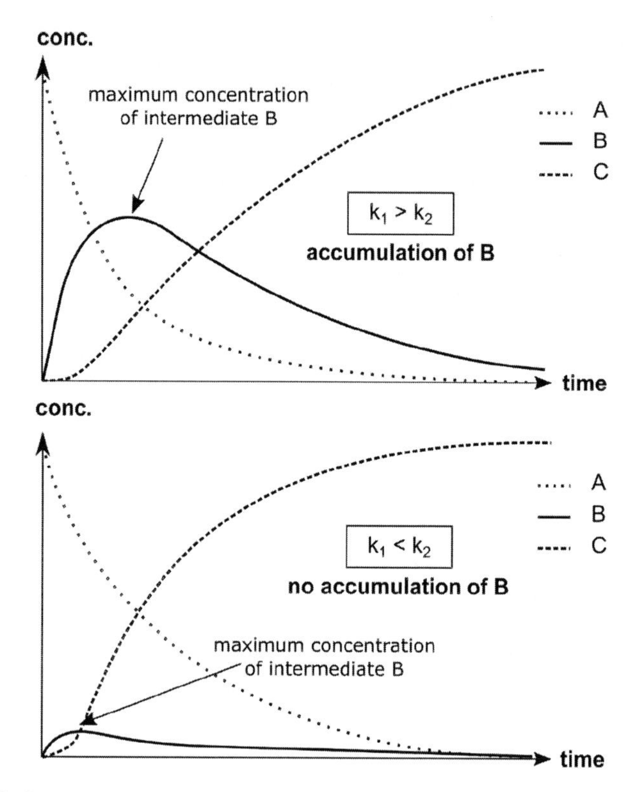

Figure 3.9. Concentration vs. time in two distinct cases. Top: $k_1 > k_2$. Bottom: $k_1 < k_2$.

This can be exploited for cross-coupling substrates that are unstable in solution if a valid precursor can be found such that the rate constant of cross-coupling substrate formation from the precursor is smaller than the cross-coupling of the substrate to the thermodynamically stable product.

Following this requirement, organotrifluoroborates hydrolyze in a step-wise fashion. Slow replacement of fluorine by hydroxy groups leads to the formation of the boronic acid *in situ*. Subsequent cross-coupling with the aryl halide avoids a build-up of boronic acid in the reaction mixture (Figure 3.10).[112] This correlates to the case of $k_1 < k_2$ (Figure

3.9, bottom). In this way, even unstable boronic acids like PCP boronic acid can be used for cross-coupling reactions.

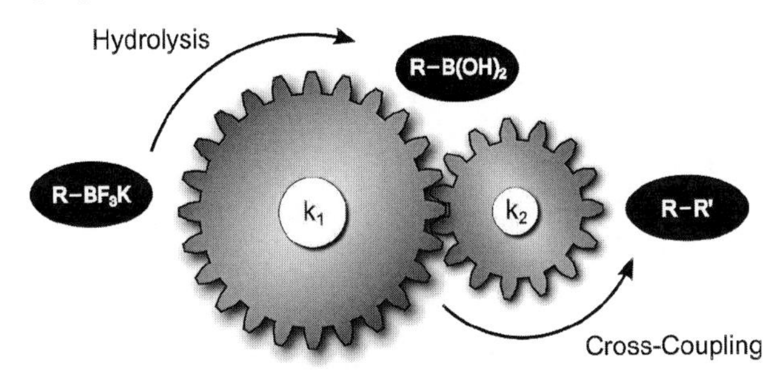

Figure 3.10 *In situ* release of the boronic acid by slow hydrolysis of the trifluoroborate and subsequent cross-coupling to the arylated PCP product.

The methodology of organotrifluoroborates nowadays widely applied in Suzuki cross-coupling chemistry[107,109,112–127] has not been used yet to circumvent the issue of unstable or unreactive PCP boron derivatives. This is however a promising approach because of the match of PCP being very electron-rich with the fact that electron-rich borates facilitate the crucial transmetalation step.[84] In the next sections, the details of synthesis and application as cross-coupling nucleophiles of PCP trifluoroborates are discussed.

3.3.2 Synthesis of PCP Trifluoroborates

Trifluoroborates are available by treatment of boronic acids or esters with the inexpensive potassium hydrogen difluoride KHF_2. Keeping in mind that PCP boronic acid **55** is unstable, it was aimed to use a one-pot procedure to generate a semi-stable PCP boronic ester *in situ* and convert it to the stable PCP trifluoroborate **62** (Scheme 3.18) without intermediate isolation similar to organotrifluoroborate syntheses described in literature.[128] The reaction starts from 4-bromo[2.2]paracyclophane (**51**) by a lithium-halogen exchange to generate the reactive intermediate 4-lithio[2.2]paracyclophane followed by electrophilic trapping with trimethyl borate to generate PCP methyl boronic ester **61**. This semi-stable boronic ester is treated with aqueous potassium hydrogen fluoride *in situ* to displace the methoxy groups by fluorides and generate the trifluorinated anionic borate **62**.

This protocol not only delivers trifluoroborate **62** in very good yield, it also allows for the large scale synthesis of **62**, which has been demonstrated by product yields of 5.00 g.

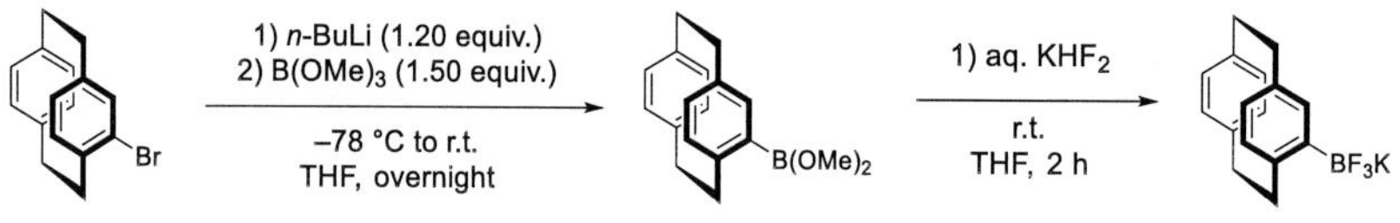

Br; 1) *n*-BuLi (1.20 equiv.) 2) $B(OMe)_3$ (1.50 equiv.); –78 °C to r.t. THF, overnight; $B(OMe)_2$; 1) aq. KHF_2; r.t. THF, 2 h; BF_3K

51 **61** **62** (87%)

Scheme 3.18. High-yielding one-pot synthesis of PCP potassium trifluoroborate **62** by *in situ* lithiation, borylation and fluorination of PCP bromide **51**.

Using the same methodology and the knowledge obtained from the selective monolithiation (see section 3.2.1.2), two more PCP trifluoroborates were accessible by this route (Scheme 3.19). The remaining bromine substituent in **49** and **50** opens the way towards further functionalization of the cross-coupled products. Therefore, these compounds are of analogous functionality as the haloboronic acids reported by Burke *et al.*[9] and I therefore suggest calling these molecules halotrifluoroborates. However, the possibility of polymerization of **49** or **50** by homocoupling has to be kept in mind when looking towards cross-coupling reactions.

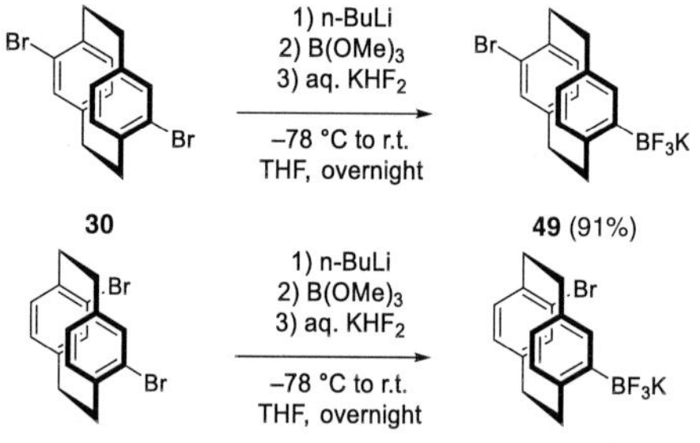

Br; Br; 1) n-BuLi 2) $B(OMe)_3$ 3) aq. KHF_2; –78 °C to r.t. THF, overnight; Br; BF_3K

30 **49** (91%)

Br; Br; 1) n-BuLi 2) $B(OMe)_3$ 3) aq. KHF_2; –78 °C to r.t. THF, overnight; Br; BF_3K

63 **50** (66%)

Scheme 3.19. Halotrifluoroborates in pseudo-para (**49**) and pseudo-ortho configuration (**50**) are accessible by the same synthetic route.

The trifluoroborate **62** and halotrifluoroborates **49** and **50** are non-toxic, free-flowing, crystalline, white solids that are long-term bench-top stable. The stability was tested by storing a sample of **62** on air for three months and observation of no signs of degradation in the ^{1}H NMR spectrum. The PCP trifluoroborates are accessible on a multi-gram scale and easily prepared from commercially available [2.2]paracyclophane in good to excellent yield. This is a significant advantage over previous PCP cross-coupling nucleophiles, which are highly reactive and have to be prepared freshly or even *in situ* for procedures like Kumada (PCP-MgBr), Negishi (PCP-ZnCl) and Stille (PCP-SnR_3) cross-coupling reactions.[88] Even worse, PCP stannanes used in Stille cross-couplings are known to be toxic[129] and their use should be avoided when possible.

3.3.3 Suzuki Cross-Coupling of PCP Trifluoroborates

3.3.3.1 Ligand-free Suzuki cross-coupling

With the PCP trifluoroborate in hand, a first set of cross-coupling test reactions was conducted. After an initial hit with Pd(II) acetate as palladium source, optimization (Table 2) was conducted with the valuable *N*-heteroarylated PCP **15** as target. The solvents most often used for trifluoroborate cross-couplings are alcohols like methanol or ethanol.[130] Due to the insolubility of PCP trifluoroborate **62** in these solvents, a mixture of toluene/water 3:1 was found to a suitable replacement. As the reaction is not very air-sensitive, reagent grade solvents in a crimp vial under argon were sufficient, omitting the need for rigorously degassed solvents as are needed in case of Stille cross-couplings. Cesium carbonate was found as the best base for the reaction and 80 °C as optimal temperature. The addition of SPhos or XPhos as common phosphine ligands to improve cross-coupling reactions mediated by $Pd(OAc)_2$ did not lead to significant improvements. With these conditions (cross-coupling procedure **3.3A**) heteroarylated PCP **15** was accessible in moderate yield (43%).[131] This is comparable to the yield of the same compound made by the Stille cross-coupling protocol.[88]

Table 2. Optimization for the synthesis of heteroarylated **15**.

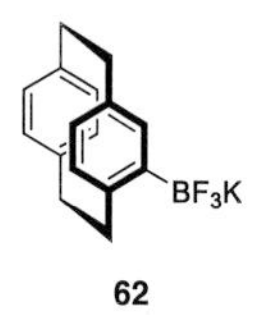

$Pd(OAc)_2$ (5 mol%)
K_2CO_3
toluene/water 3:1
80 °C, overnight
64
15

Entry	deviation from standard conditions	Yield[a] [%]
1	none	27
2	48 h	37
3	15 mol% SPhos	25
4	15 mol% XPhos	35
5	100 °C	22
6	K_3PO_4	30
7	**Cs_2CO_3**	**43**
8	KOH	31
9	KF	9
10	Cs_2CO_3, 72 h	32
11	10 mol% catalyst	45
12	2 mol% catalyst	37
13	1 mol% catalyst	27

[a] NMR yields with 1,3,5-trimethoxybenzene as internal standard

However, with these conditions, only the substrate scope shown in Scheme 3.20 could be realized. The cross-coupling of PCP with 2-bromopyridine was reported to be challenging and low-yielding.[88] Thus, it is not surprising that the yields for the other substrates are much more satisfying. It could be shown that this cross-coupling protocol is suited for larger scale reactions, as bipyrimidine **67** was obtained on a 4 mmol scale without any reduction in yield.

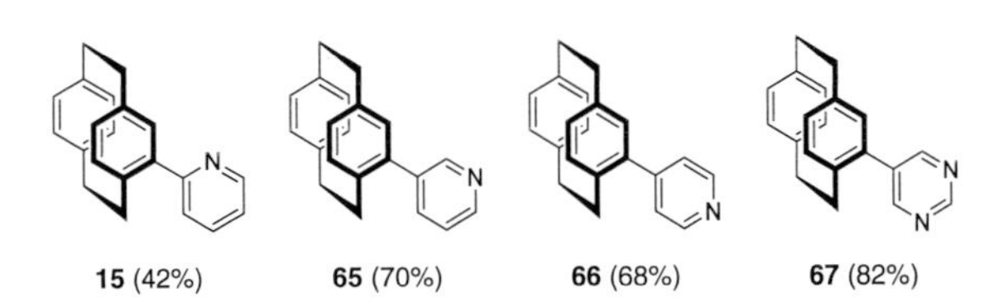

Scheme 3.20. Substrate scope for the ligand-free cross-coupling of trifluoroborate **62** towards *N*-heteroarylated PCPs (**15, 65-67**) using general cross-coupling procedure **3.3A**.

All efforts to use non-*N*-heteroaryl bromides (e.g. phenyl, tolyl and methoxyphenyl bromides) were met with failure, as an analysis of the reaction products showed almost complete conversion of trifluoroborate **62** to the protodeboronation product (PCP) along with the oxidized 4-hydroxy[2.2]paracyclophane **56**.[131]

3.3.3.2 RuPhos mediated Suzuki cross-coupling reactions

Given the success of the previous cross-coupling procedure, it was tried to render the cross-coupling of non-heteroaryl bromides successful. However, SPhos and XPhos were not able to provide the necessary reactivity. Given PCPs less-than-aromatic character, it was tested whether RuPhos, a phosphine ligand known for promoting successful cross-coupling of alkyl trifluoroborates, could also be applied for this cross-coupling endeavor. The choice of this phosphine ligand turned out to be crucial. The sensitivity towards choice of the phosphine ligand or "fine-tuning" of the reaction conditions can be appreciated with a look at the structures of the three phosphine ligands tested (Figure 3.11). Although very minor in structural difference, the outcome of the reaction (detection of product by GCMS) was affected dramatically. This is a very common finding in palladium-catalyzed cross-couplings. For every reaction class, a specialized ligand system has to be found to yield optimal results.[83] Bulky, electron-rich phosphine ligands are thought to increase the efficiency of palladium-catalyzed cross-couplings through (i) donation of electron-density to the intermediate Pd(0) complex to facilitate oxidative addition and (ii) formation of highly reactive L_1Pd species because of ligand-ligand repulsion forces.[83]

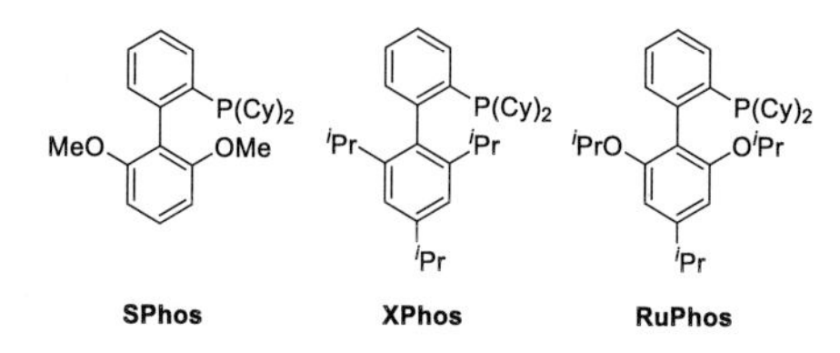

Figure 3.11. Chemical structures of three common phosphine ligands: SPhos, XPhos and RuPhos.

After reaction optimization, the following substrate scope could be achieved with the conditions (general cross-coupling procedure **3.3B**) given in Table 3, delivering good to excellent yields for a very wide range of aryl bromides, chlorides and triflates. Electron withdrawing (entries 2-6) as well as electron donating groups gave satisfying yields. A range of functional groups such as esters, nitriles, ketones, alkynes, ethers, nitroarenes and amines were tolerated. Steric hindrance diminished the yield of the reaction (entry 15). Not only arenes but also benzylic bromides were accessible by this methodology. It was not possible to couple sp^3-hybridized bromocyclohexane (entry 17). Also, aromatic amines proved to be challenging substrates (entry 9), delivering only moderate yields.

Table 3. Substrate scope for the RuPhos-mediated Suzuki cross-coupling reaction of trifluoroborate **62** using general cross-coupling procedure **3.3B.**

BF_3K **62** → Ar **15,68-82**

$Pd(OAc)_2$, **RuPhos**, K_3PO_4
Tol/water 10:1, 80 °C, overnight
Ar–X
X = Cl, Br, OTf

Entry	Ar	Product	Yield[a] [%]
1	N	15	39[b]
2	O, OMe	68	92[b]
3	O, Me	69	81[b], 69[c], 79[d]
4	NO_2	70	87[b]

5	F, –F, F	71	74[b]
6	–CN	72	92[b]
7	–Me	73	91[b]
8	–OMe	74	86[b]
9	–NH_2	75	23[b]
10	–N	76	88[b]
11	N=	77	90[b]
12	–TMS	78	78[b]
13		79	34[b]
14	S	80	87[b]
15	Me, –Me, Me	81	25[b]
16	–CN	82	51[b]
17		83	-

[a]Isolated yields, [b]aryl bromide, [c]aryl chloride, [d]aryl triflate.

Despite these good results, the synthesis of **15** (Table 3, entry 1) was still lacking satisfying yields. After RuPhos had been discovered as the ligand of choice for these transformations, another optimization was conducted (Table 4) to find better conditions tailored to the synthesis of **15**. With these new conditions, a final yield of 71% could be achieved, which is a remarkable improvement over previous methods to synthesize this compound. This work was done within a bachelor thesis project carried out by Jannik

Schlindwein.[132] It is speculated that the improved yields are largely dependent on (i) the weaker base sodium carbonate compared to the other entries and (ii) the lower concentration of sodium carbonate in the aqueous phase when using 1:1 toluene/water as a solvent. Both of these conditions should lead to a slower hydrolysis of **62** which in turn compensates for the slow rate of the cross-coupling with the sluggish cross-coupling partner **64**, thus avoiding build-up and decomposition of the intermediate boronic acid.

Table 4. Optimization for the RuPhos mediated synthesis of 2-pyridyl PCP **15**.

$Pd(OAc)_2$ (10 mol%)
RuPhos
K_3PO_4
Tol/H_2O 10:1 (0.1 M)
80 °C, overnight

BF_3K

N Br

N

62 **64** **15**

Entry	Deviation from standard	Yield[a] [%]
1	none	39
2	Tol/H_2O 3:1	41
3	Tol/H_2O 1:1	48
4	Tol/H_2O 1:1, K_2CO_3	63
5	Tol/H_2O 1:1, KOH	58
6	Tol/H_2O 1:1, Cs_2CO_3	61
7	**Tol/H_2O 1:1, Na_2CO_3**	**71**
8	Tol/H_2O 1:1, Na_2CO_3, 60 °C	56
9	Tol/H_2O 1:1, Na_2CO_3, 100 °C	42
10	Tol/H_2O 1:1, Na_2CO_3, 0.2 M	57
11	Tol/H_2O 1:1, Na_2CO_3, 2 mol% cat	66
12	Tol/H_2O 1:1, Na_2CO_3, 5 mol% cat	73
13	Tol/H_2O 1:1, Na_2CO_3, 1 mol% cat	48
14	Tol/H_2O 1:2, Na_2CO_3	70
15	Tol/H_2O 1:5, Na_2CO_3	71

[a]Yields determined by ^1H NMR, 1,3,5-trimethoxybenzene as internal standard was used.

Additionally, the reaction of trifluoroborate **62** with PCP bromide **55** was probed to generate dimer **84** (Scheme 3.21). The product was not observed, which could be possible explanation by the high steric repulsion between the two PCP units that has to be overcome to form the new carbon–carbon bond.

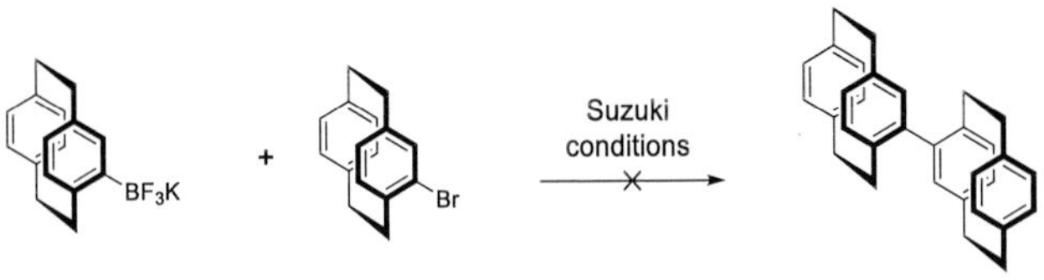

Suzuki conditions

62 55 84

Scheme 3.21. Unsuccessful reaction to generate PCP dimer **84**. Only starting material **55** and the protodeboronation product of **62** were found.

3.3.3.3 Suzuki Cross-Coupling of Halotrifluoroborates

The halotrifluoroborates **49** and **50** are valuable intermediates on the way towards heterodisubstituted PCPs. Since it was known that a homocoupling of **49** does not happen, cross-coupling with other substrates is possible. A first test reaction (Table 5, entry 1) showed poor yield for the cross-coupling of 4-bromo(2'pyridyl)benzene (**85**) with halotrifluoroborate **49**. Thus, a further optimization study was conducted with results summarized in Table 5. Interestingly, an increase of the stoichiometric amount of the trifluoroborate salt led to a decrease in yield. The most important aspect turned out to be the concentration of the reagents in the organic phase, with 0.2 M being the sweet spot for a satisfying yield of 50%. This could be due to the rate of catalytic reactions being highly dependent on the concentration of the reagents in solution. This is detrimental here, as the boronic acid released by hydrolysis from halotrifluoroborate **49** is not stable, thus the subsequent cross-coupling step has to be of sufficient rate to generate the desired product instead of the protodeboronated 4-bromo[2.2]paracyclophane.

Br BF$_3$K + N Br Pd(OAc)$_2$ RuPhos K$_3$PO$_4$ Tol/H$_2$O (10:1, 0.1 M) 80 °C, overnight Br N

49 85 86 (22%)

Scheme 3.22. Suzuki cross-coupling of halotrifluoroborate **49** with **85** to yield heterodisubstituted PCP **86**. Conditions before optimization.

Table 5. Optimization for the RuPhos mediated cross-coupling synthesis of **86**.

Br BF$_3$K Pd(OAc)$_2$ RuPhos K$_3$PO$_4$ Tol/H$_2$O (10:1, 0.1 M) 80 °C, overnight N Br Br N

49 85 86

Entry	Deviation from standard	Yield [%][a]
1	none	22
2	1.00 equiv. **49**	35
3	2.00 equiv. **49**	18
4	Tol/H_2O 10:1, 0.2 M	50
5	Tol/H_2O 1:1, 0.2 M	50
6	Tol/H_2O 10:1, 0.6 M	37
7	Na_2CO_3, 0.2 M	16
8	K_2CO_3, 0.2 M	43
9	Cs_2CO_3, 0.2 MS	44

[a]Yields were determined by ^{1}H NMR, 1,3,5-trimethoxybenzene was used as internal standard.

3.3.4 Conclusions

The chemistry of PCP has been lacking a versatile, inexpensive, safe, convenient and modular Suzuki cross-coupling protocol. This could be achieved by applying the methodology of trifluoroborates as boronic acid precursors. Aryl bromides, chlorides and triflates bearing a very wide range of functional groups could be cross-coupled with the PCP trifluoroborates in moderate to excellent yields. In its convenience and versatility, this reaction is superior to any previously reported cross-coupling methods of PCP. It is suited to accelerated generation of PCP containing chemical libraries as the PCP core can be introduced in almost any substrate bearing an aryl halide or pseudohalide in its structure. All these advantages are further emphasized by the ease of access on a multigram scale and indefinite storability on the bench top of the trifluoroborates.

3.4 PCP-Porphyrin Conjugates

3.4.1 Introduction to Porphyrins

Porphyrins are heterocyclic macrocycles that are derived from the parent structure porphin **87** (Scheme 3.23). While porphin itself is more of theoretical interest, it is the parent compound of the class of porphyrins. These molecules are called free-base porphyrins when they are not coordinated to a metal. Porphyrins, and especially their metal complexes, are very common in nature and can be found as hemes[133] in animals, chlorophylls in plants[134] and in the mineral abelsonite,[135] to name a few.

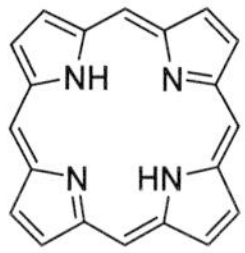

NH N
N HN

87

Scheme 3.23. Porphin, the parent compound of porphyrins.

Due to the planar, extended π-system in porphyrins, they are considered aromatic compounds and absorb strongly in the visible region, thus rendering these molecules intensely colorful. Even their name itself is derived from the Greek word *porphyros* for *purple*.[136] They found extensive application in fields such as photodynamic therapy,[137] supramolecular chemistry[138,139] and biomimetic catalysis.[140]

Besides their technical applications, they have been named the "ligands par excellence of biology",[141] paying credit to their ability to form complexes with a wide range of transition metals and, thus supporting life.

The laboratory synthesis of porphyrins is usually performed by condensation of suitable pyrroles and aldehydes, initially developed by Rothemund,[142] improved by Adler and Longo,[143] and finally by Lindsey *et al.*, working in high dilution.[144–146]

In this work, porphyrin's ability to be easily modified and to form stable transition metal complexes will be exploited to access heterobimetallic complexes.

3.4.2 Synthetic Access to Porphyrins on PCP

3.4.2.1 Cross-Coupling Access to PCP-Porphyrins

Porphyrins, if adequately substituted, can be introduced to the PCP by means of cross-coupling chemistry. Different types of cross-coupling reactions were successfully employed in the past and are presented in the following.

3.4.2.1.1 Stille Cross-Coupling[147]

Using organotin reagents, both free-base and metallated (M = Cu, Zn) porphyrins are valid substrates for Stille cross-coupling. The underlying protocol has been published by Braun *et al.* and could successfully be employed in the synthesis of PCP-porphyrins **92-94**.

CuBr
LiCl
$Pd(PPh_3)_4$
DMF
100 °C, 3 d

88

M = 2H **89**
M = Cu **90**
M = Zn **91**

M = 2H **92**
M = Cu **93** (52%)
M = Zn **94** (48%)

Scheme 3.24. Synthesis of PCP-porphyrin conjugates **92-94**.[147]

Interestingly, the free-base product **92** could not be detected and instead, the copper complex **93** was found. This could be due to the additive Cu(I) which contains a Cu(II) impurity, as well as unwanted oxidation reactions that can generate Cu(II) from Cu(I). The low yield of this reaction could be dramatically improved by using the already copper metallated **90** and led to the desired copper or zinc complex **93** or **94** in 52% and 48% yield, respectively.

The pseudo-*para* disubstituted variant of the zinc complex was aimed for by using either the 4,16-dibromoPCP **30** and coupling with the organotin version of the porphyrin **96** or the organotin PCP **95** with the bromoporphyrin **91**. However, the organotin PCP did afford the product in only 2% yield (path b, Scheme 3.25), whereas the inverse reaction delivered it in 30% yield (path a).

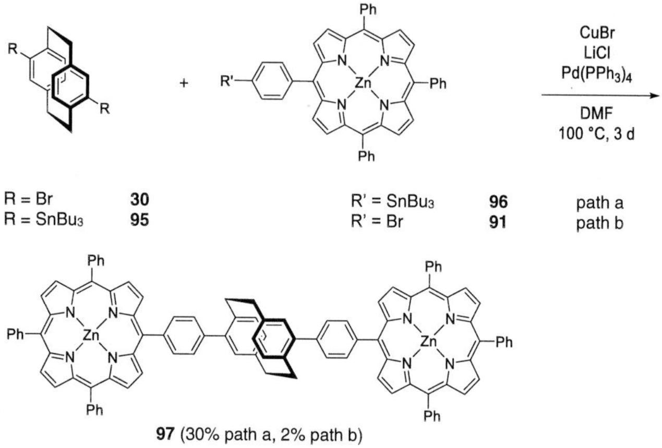

Scheme 3.25 shows the coupling of PCP **30** (R = Br) or **95** ($R = SnBu_3$) with zinc porphyrin **96** ($R' = SnBu_3$, path a) or **91** ($R' = Br$, path b), using CuBr, LiCl, $Pd(PPh_3)_4$ in DMF at 100 °C for 3 d, to give **97** (30% path a, 2% path b).

Scheme 3.25. Synthesis of homobimetallic PCP-porphyrin conjugate **97**.[147]

3.4.2.2 Condensation Access to PCP-Porphyrins

As described in the introductory section, condensation of suitable pyrroles and aldehydes leads to porphyrins. As the aldehyde of PCP is easily accessible and additionally shows several interesting advantages (see section 3.5.2), mixed condensation with pyrroles or precondensed dipyrromethanes should lead to PCP-porphyrin conjugates as well.

The conditions optimized by Christoph Schissler of the Bräse group could be applied to generate free-base PCP-porphyrin conjugates with different substitution patterns. A schematic reaction overview is shown in Scheme 3.26.

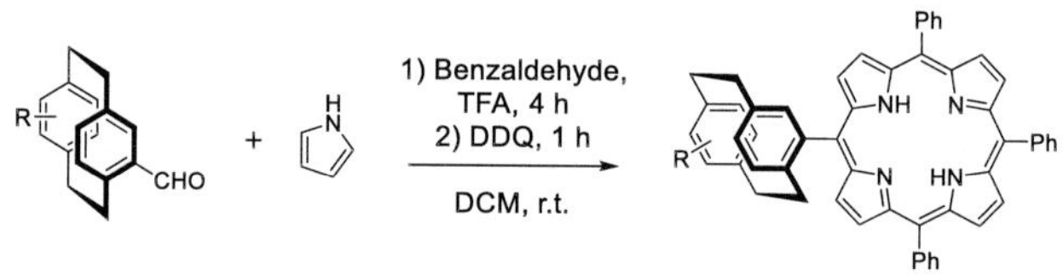

Scheme 3.26 shows the condensation of PCP aldehyde **98** with pyrrole **99** (1) Benzaldehyde, TFA, 4 h; 2) DDQ, 1 h; DCM, r.t.) to give **100**.

Scheme 3.26. Synthesis of PCP-porphyrin conjugates **100** by a condensation reaction.[148]

3.4.3 Heterobimetallic Complexes

3.4.3.1 Zn/Au PCP-Porphyrin Conjugates

To generate heterobimetallic complexes with PCP-porphyrin conjugates, it was first tried to prepare a bromine functionalized PCP-porphyrin to introduce a second coordination site by lithium-halogen exchange reactions. It was envisioned to transform the bromine handle to a phosphinyl binding site able to coordinate a gold(I) atom as demonstrated in example Zn/Au heterobimetallic complex **101**.

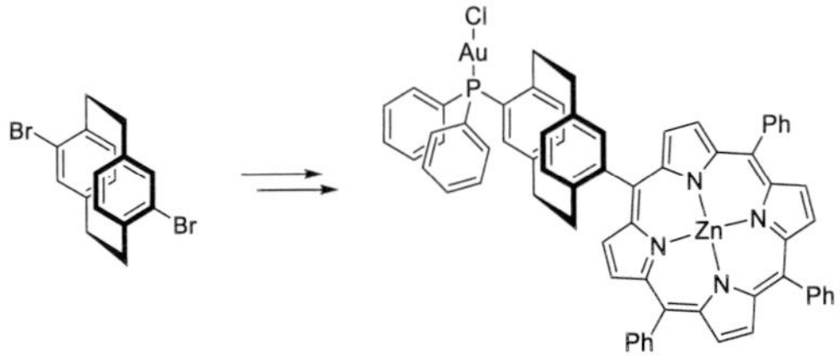

45 **101**

Scheme 3.27. Synthesis of zinc PCP-porphyrin conjugate **102** by condensation and metalation.[148]

To this end, 4-bromo-16-formyl[2.2]paracyclophane (**45**) was condensed with pyrrole and benzaldehyde to generate the free-base bromo substituted PCP-porphyrin conjugate, which after metalation afforded zinc-porphyrin **102** (Scheme 3.28).

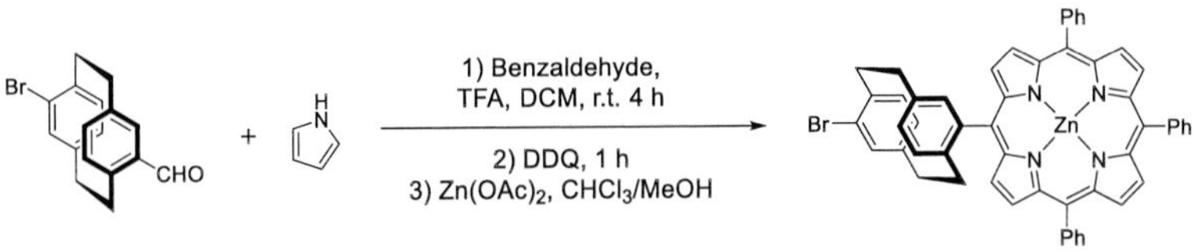

45 **99** **102** (11% over 2 steps)

Scheme 3.28. Synthesis of zinc PCP-porphyrin conjugate **102** by condensation and metalation.[148]

It was then tried to subject **102** to a lithium-halogen exchange and subsequent phosphination to access phosphine/porphyrin PCP **103** (Scheme 3.31). However, even after treating **102** with a large excess of *n*-BuLi or *t*-BuLi mostly starting material remained. While not decomposing, **102** seems to almost completely withstand the lithium-halogen exchange reaction conditions.

1) *n*-BuLi or *t*-BuLi, –78 °C, 30 min
2) PPh_2Cl, –78 °C to r.t., 2 h

102 **103**

Scheme 3.29. Unsuccessful lithiation to access phosphine/porphyrin PCP **103**.

The strategy was changed accordingly for the more challenging pseudo-*ortho* substitution pattern and additionally avoiding the synthetic bromine handle. This was achieved by conversion of the bromine to a TMS protected alkyne **104** by a Sonogashira cross-coupling reaction before porphyrin condensation (Scheme 3.30).

TMS-acetylene
$Pd(PPh_3)_2Cl_2$
PPh_3
CuI
THF/NEt_3
75 °C, overnight

46 **104** (52%)

Scheme 3.30. Conversion of the bromine to a protected alkyne group prior to condensation to form the porphyrin.

The aldehyde **104** was subsequently condensed and metallated to yield the porphyrin zinc complex **105** (Scheme 3.31).

1) Benzaldehyde, TFA, DCM, r.t. 4 h
2) DDQ, 1 h
3) $Zn(OAc)_2$, $CHCl_3$/MeOH

104 **105** (3%)

Scheme 3.31. Conversion of the aldehyde **104** to the porphyrin zinc complex **105**.[148]

Alkynes can form linear complexes with suitable gold(I) precursors after deprotection with a fluoride source and deprotonation by a base. This was aimed for with the porphyrin zinc complex **105**. The complex **105** was subjected to deprotection conditions with TBAF to remove the TMS group and deliver the free alkyne **106** in excellent yield (Scheme 3.32).

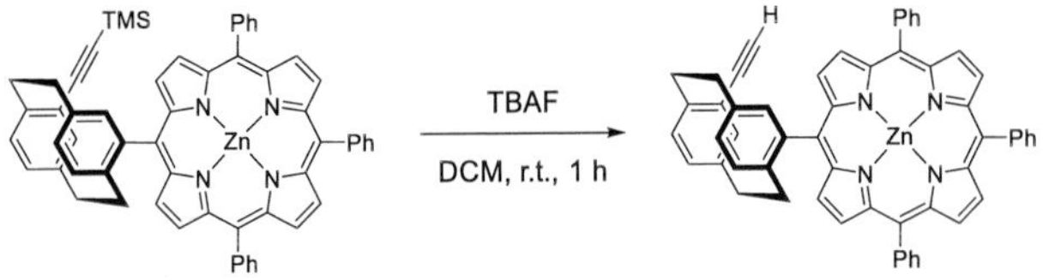

TBAF

DCM, r.t., 1 h

105 **106** (99%)

Scheme 3.32. Deprotection of the TMS protecting group to afford the free alkyne **106**.[148]

The next step was the deprotonation and metallation to introduce gold(I) as a second metal to the complex. First, the metallation precursor **108** was made from tetrahydrothiophene gold(I) chloride and triphenylphosphine (Scheme 3.33) in good yield.

PPh_3

DCM, r.t. overnight

107 **108** (78%)

Scheme 3.33. Preparation of gold(I) precursor **108**.

Finally, the zinc complex **106** was deprotonated with potassium hydroxide and metallated *in situ* with the precursor **108**. However, purification of the gold(I)-zinc(II) heterobimetallic complex **109** by means of crystallization or silica chromatography did not lead to the pure compound. Formation of the complex could be verified by ESI MS (Figure 3.12), but its stability seems to be not sufficient for purification, as only traces of the complex were found in ESI MS after chromatography purification attempts.

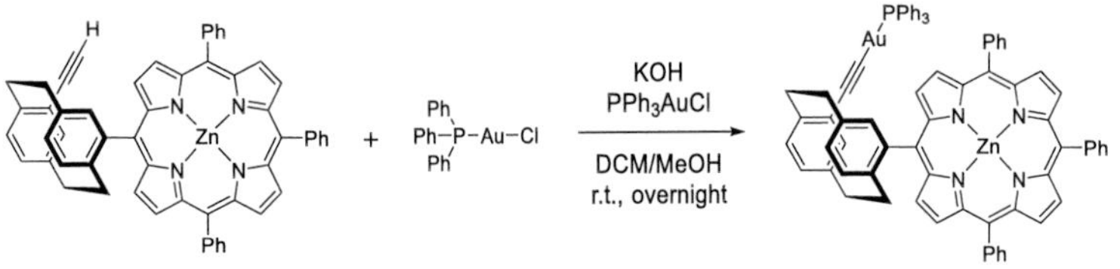

106 108 109 (not isolated)

Scheme 3.34. Auration of the free alkyne **106** with gold(I) precursor **108** after deprotection by potassium hydroxide to afford heterobimetallic complex **109**.

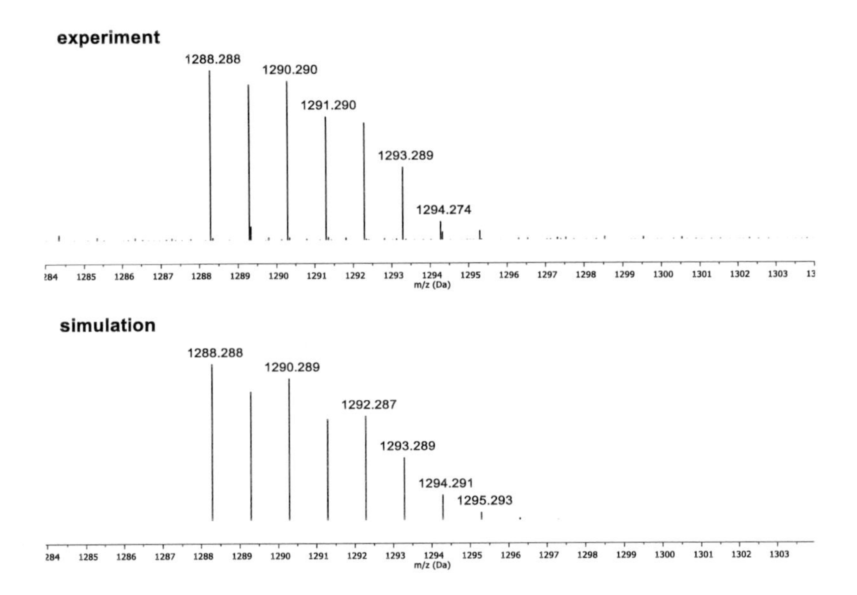

Figure 3.12. Experiment vs. simulation ESI MS of gold(I) zinc(II) PCP-porphyrin conjugate complex **109**.

3.4.3.2 Pseudo-*para* Zn/Ru PCP-Porphyrin Conjugates

After this, it became clear that a different ligand-metal combination should be tried with the PCP-porphyrin motif, specifically one that is more stable than the labile alkyne gold complexes.

To this end, pseudo-*para* formyl/phenylpyridine disubstituted PCP **111** was synthesized. The selective monolithiation protocol could be applied to obtain formyl/bromo PCP **45** (Scheme 3.35).

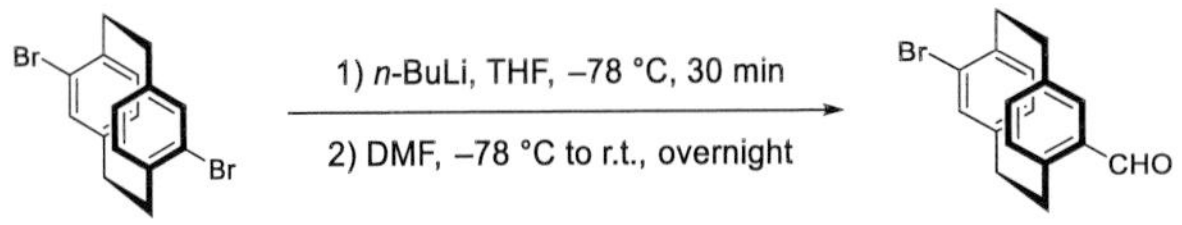

1) *n*-BuLi, THF, –78 °C, 30 min
2) DMF, –78 °C to r.t., overnight

30 **45** (75%)

Scheme 3.35. Selective monolithiation followed by formylation with DMF to afford 4-bromo-16-formyl[2.2]paracyclophane **45**.

The remaining bromine handle could be used for a Suzuki cross-coupling to install the phenylpyridine moiety to ultimately obtain formyl/phenylpyridine PCP **111** (Scheme 3.36).

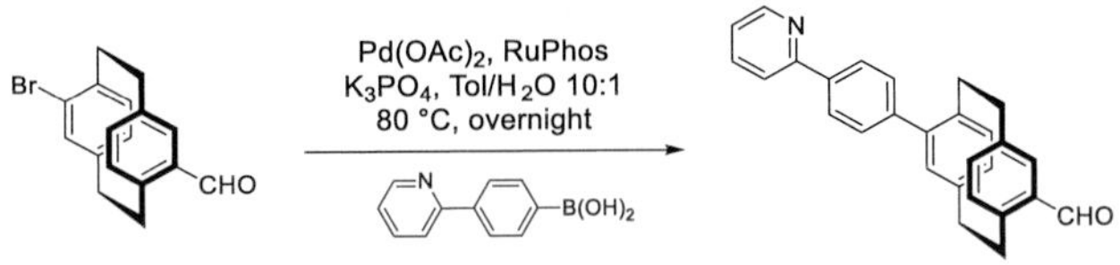

$Pd(OAc)_2$, RuPhos
K_3PO_4, Tol/H_2O 10:1
80 °C, overnight

30 **110** **111** (11%)

Scheme 3.36. Suzuki cross-coupling to obtain formyl/phenylpyridine PCP **111**.

Finally, the formyl substituent was used for the *de novo* synthesis of the porphyrin ring by a condensation reaction with benzaldehyde and pyrrole (Scheme 3.37). The free-base porphyrin **112** was obtained in 6% yield.

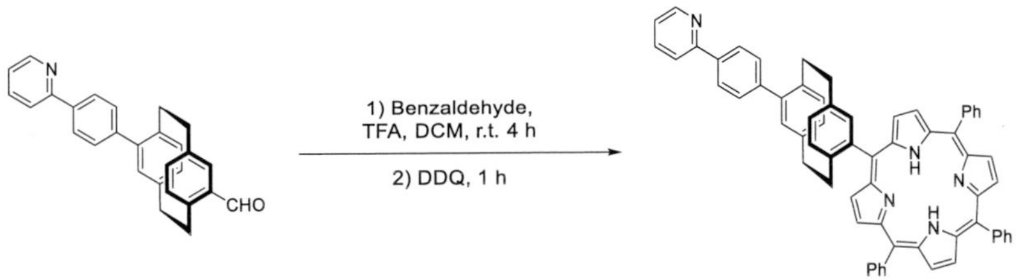

1) Benzaldehyde, TFA, DCM, r.t. 4 h
2) DDQ, 1 h

111 **112** (6%)

Scheme 3.37. Porphyrin condensation to afford PCP-porphyrin conjugate **112**.[148]

Next, the first metalation step was conducted with standard conditions for the insertion of zinc into the porphyrin moiety (Scheme 3.38).[148] The product **113** was obtained in quantitative yield and subjected to the ruthenation in the next step.

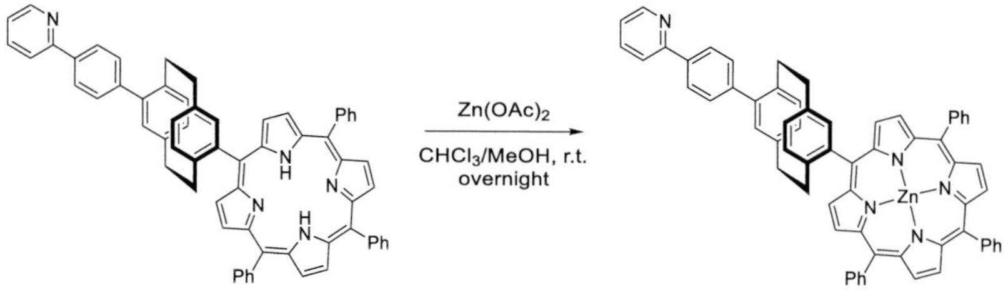
N
N
H
N
N
H
N
Ph
Ph
Ph
Zn(OAc)2
CHCl3/MeOH, r.t.
overnight
N
N
N
Zn
N
N
Ph
Ph
Ph

112

113 (quant.)

Scheme 3.38. Zinc metalation of free-base **112** to afford phenylpyridine substituted PCP-porphyrin conjugate zinc complex **113**.[148]

Standard ruthenation conditions[149] were applied to achieve cycloruthenation of the free phenylpyridine binding site (Scheme 3.39). The initial use of 1.00 equiv. ruthenium per binding site led to almost no conversion as determined by TLC. The reaction was thus repeated with 10.0 equiv. ruthenium per binding site to afford the heterobimetallic PCP-porphyrin conjugate **114**. The simulated ESI MS data agrees very well with the recorded data (Figure 3.13). However, the obtained mixture was not separable by column chromatography.

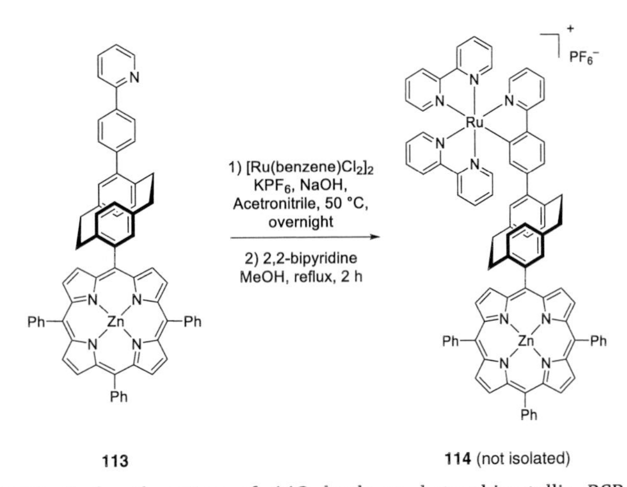

Scheme 3.39. Cycloruthenation of **113** leads to heterobimetallic PCP-porphyrin conjugate **114**.

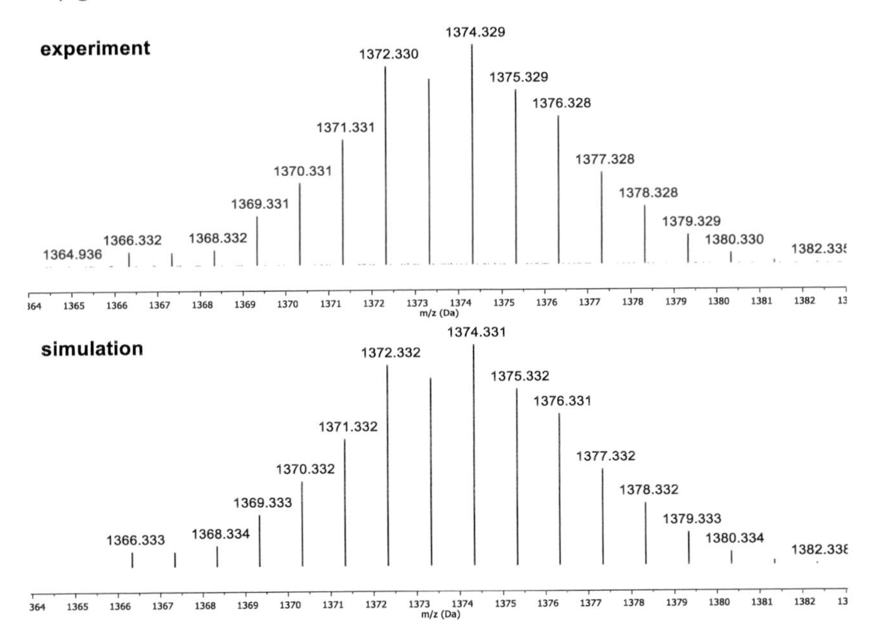

Figure 3.13. Experimental vs. simulated positive mode ESI mass spectrum of **114**.

3.4.3.3 Pseudo-*ortho* Zn/Ru PCP-Porphyrin Conjugates

Next, it was planned to investigate the same methodology for access towards the pseudo-*ortho* analogue of **114**.

Starting from pseudo-*ortho* dibromo PCP **34** which is readily available by thermal isomerization,[150] the formyl/bromo PCP **46** was accessed in good yield (Scheme 3.35).

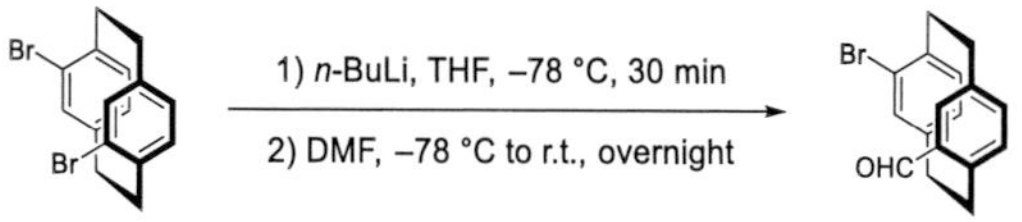

1) *n*-BuLi, THF, –78 °C, 30 min
2) DMF, –78 °C to r.t., overnight

34 **46** (56%)

Scheme 3.40. Selective monolithiation followed by formylation with DMF to afford 4-bromo-12-formyl[2.2]paracyclophane **46**.

Following, a Suzuki cross-coupling with phenylpyridine boronic acid **110** afforded the desired pseudo-*ortho* phenylpyridine/formyl PCP **115** (Scheme 3.41).

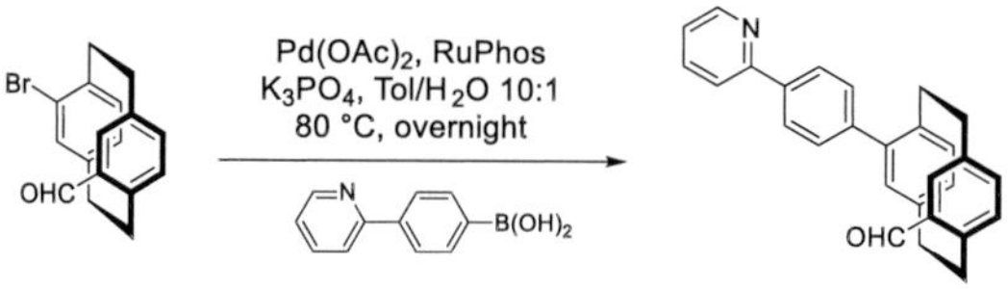

$Pd(OAc)_2$, RuPhos
K_3PO_4, Tol/H_2O 10:1
80 °C, overnight

46 **110** **115** (18%)

Scheme 3.41. Selective monolithiation followed by formylation with DMF to afford phenylpyridine/formyl PCP **115**.

Subsequent condensation with pyrrole and benzaldehyde led to the *de novo* synthesis of the porphyrin ring in pseudo-*ortho* PCP-porphyrin conjugate **116** (Scheme 3.42).

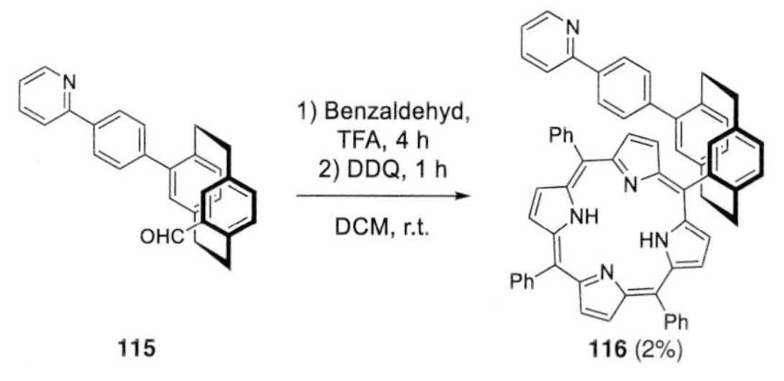

Scheme 3.42. Porphyrin condensation synthesis to afford pseudo-*ortho* PCP-porphyrin conjugate **116**.[148]

Again, the first metalation was done with standard zinc metalation conditions to obtain the Zn complex **117** (Scheme 3.43). However, the zinc metalation led to a mixture that was not separable by column chromatography. The product was found in ESI MS.

$Zn(OAc)_2$

$CHCl_3$/MeOH, r.t.
overnight

116

117 (not isolated)

Scheme 3.43. Zinc metalation to afford pseudo-*ortho* PCP-porphyrin conjugate zinc complex **117**.[148]

3.4.4 Conclusions

It was shown that PCP-porphyrin conjugates show very high potential for distance-controlled heterobimetallic complexes. The remarkable stability of porphyrin metal complexes is of great advantage for this and is complemented by porphyrin's other qualities like absorption characteristics, planar geometry and aromaticity. The initially explored cross-coupling paths only delivered homobimetallic complexes, as metalation of the porphyrin was necessary for the cross-coupling to be successful. Classical condensation reactions for the *de novo* synthesis of the porphyrin scaffold proved much more versatile. However, thorough screenings showed that the porphyrin substituent on the PCP made subsequent transformation of the remaining bromine synthetic handle on the PCP challenging. Thus, it was required to first functionalize the second handle and make use of an aldehyde synthetic handle on PCP for a subsequent porphyrin *de novo* synthesis by condensation with pyrrole and aromatic aldehydes. In this way, a range of Zn/Au and Zn/Ru heterobimetallic complexes were accessible. While the Zn/Au complexes turned out to be difficult to handle and isolate, the cycloruthenated Zn/Ru PCP-porphyrin conjugate complexes showed great potential for future investigations into this class of compounds. The late metalation step in the synthesis allows for a range of metals to be inserted selectively into the porphyrin and opens the door to the preparation of a chemical library of diverse heterobimetallic PCP-porphyrin conjugate complexes. However, separation of the obtained mixtures by column chromatography remains challenging.

3.5 Cyclometallated 2-Oxazolinyl PCP complexes

3.5.1 Introduction

In pursuit of heterobimetallic complexes, it was decided to design a system having two coordination sites with the PCP as linking backbone between them. As shown in section 3.1.1.4, phosphines are a very good match for linear gold(I) complexes, thus one coordination site was set to be a phosphine. The other site could be made up of any known ligand motif if it selectively binds to another metal ion and does not interact with the gold(I) complex. As gold(I) does not undergo cyclometallation or bind tightly to *N*-donor ligands, cyclometallating C–N ligands are of special interest to avoid interference with pre-installed the gold(I) atom.

Recently, a series of cyclometallated ruthenium complexes containing the PCP scaffold were reported (Scheme 3.44).[63] These complexes were sufficiently stable to oxygen and moisture to be purified by column chromatography on silica, which was an important prerequisite for this work. While the neutral complexes were not accessible in good yield, addition of potassium hexafluorophosphate afforded the cationic complexes. Thus, it was decided to use the motif of cationic cyclometallated PCP complexes (**118**) to introduce ruthenium(II) as one of two metal centers. Both 2'-pyridyl and 2'-oxazolinyl PCP ruthenium(II) complexes are known. Since the latter show a better stability against organolithium reagents, which are needed to introduce the second metal, it was decided to prepare a 2'-oxazolinyl PCP.

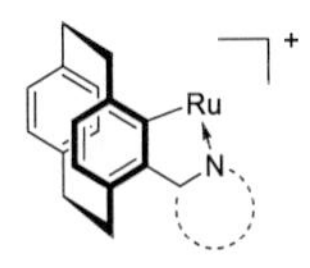

118

Scheme 3.44. Schematic depiction of a cyclometallated cationic ruthenium(II) complex with PCP as backbone.

There are several reports concerning the synthesis of PCP oxazoline derivatives.[38,74,102,150,151] All of these use common condensation reaction conditions, starting from the PCP carboxylic acid **119** and subsequently employ an Appel-like cyclization reaction to form the desired oxazoline (Scheme 3.45).

1) $SOCl_2$

2) R, H_2N, OH

R = alkyl

NEt_3, PPh_3

CCl_4

COOH

119 **120** **121**

(90% over two steps)

Scheme 3.45. Previously reported synthesis of oxazolinyl PCPs. Intermediate β-hydroxylamide **120** is subjected to the Appel-like cyclization conditions to afford 2-oxazolinyl **121**.

3.5.2 Oxazoline Synthesis from PCP Aldehyde

While the established route to access the oxazoline scaffold *via* the carboxylic acid route delivers good yields, 4-bromo-16-carboxy[2.2]paracyclophane (**48**) shows dramatically reduced solubility in all solvents tested in comparison to the non-brominated PCP **119** (Scheme 3.46). The solubility is reduced to a degree where even esterification of the carboxylic acid group becomes impossible which is necessary for purifying this intermediate.[64]

48 **119**

Scheme 3.46. PCP carboxylic acids with (**48**) and without (**119**) a further bromo-substituent.

Thus, changing the synthesis route to starting the oxazoline synthesis from PCP aldehyde **122** has multiple advantages: (i) the aldehyde shows significantly increased solubility in all solvents compared to the carboxylic acid (ii) the aldehyde offers easy access to enantiopure PCP derivatives in large scale following a procedure by Bräse *et al.*[152] (iii) a one-pot reaction to access a library of substituted 2-oxazolines is highly convenient (iv) by exchanging the amino alcohol for a diamine, imidazolines become accessible, broadening the scope of this reaction (Scheme 3.47).[153–156]

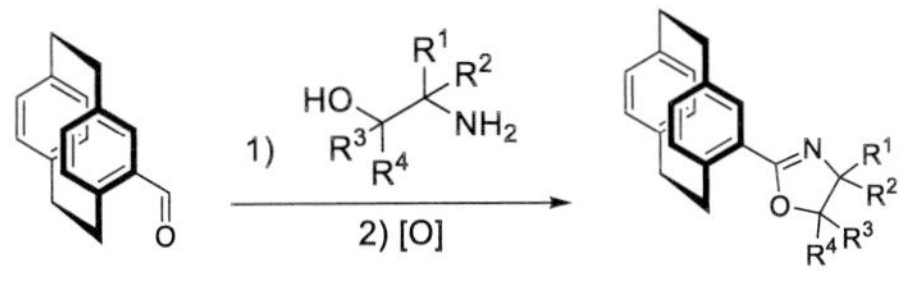

Scheme 3.47. Oxidative two-step one-pot direct conversion of PCP aldehyde **123** to PCP 2-oxazolines **124**.

Within PCP oxazolines **124**, if $R^1 \neq R^2$ or $R^3 \neq R^4$, an additional stereogenic center is generated at the oxazoline. In conjunction with the planar chirality of the PCP core, this gives rise to diastereomers. In some cases, these are separable by column chromatography on regular silica. However, to avoid introduction of more complexity, it was decided to aim for the unsubstituted parent compound, the 2-oxazolinyl substituent where $R^{1\text{-}4}$ = H. The mechanism of the oxidative conversion of PCP aldehyde to PCP 2-oxazolinyl proceeds by a condensation reaction to form β-hydroxyimine **126**.[153] Nucleophilic attack of the alcohol group at the imine carbon leads to oxazolidine **127**. In a second step, *N*-bromosuccinimide is used to brominate the nitrogen atom of the oxazolidine to afford *N*-bromo oxazolidine **128**. Spontaneous elimination of hydrogen bromide finally furnishes the double bond and results in the formation of 2-oxazolinyl[2.2]paracyclophane (**129**) in good yield.

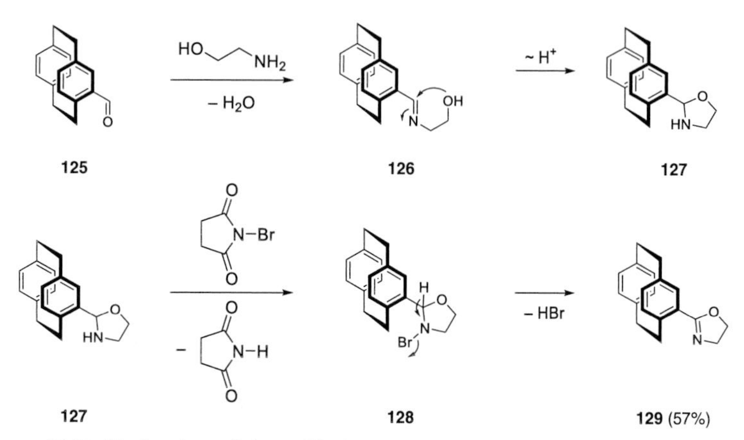

Scheme 3.48. Mechanism of the oxidative two-step one-pot direct conversion of PCP aldehyde 45 to PCP 2-oxazoline **129**.

3.5.3 Heterobimetallic Complexes of Gold and Ruthenium

To demonstrate the feasibility of the overall approach used herein, the preparation of a ditopic ligand with subsequent stepwise metalation was envisioned. To this end, cationic complex **130** containing a gold(I) and ruthenium(II) center was set as a proof-of-principle target (Scheme 3.49).

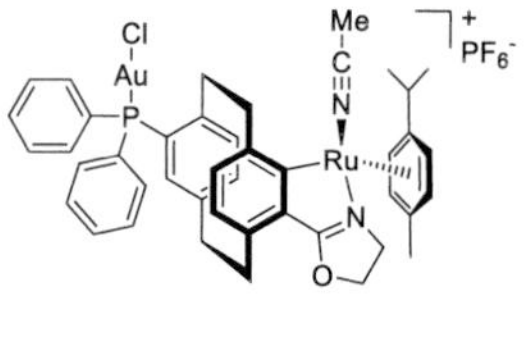

Cl Au P Me C N Ru N O PF_6^-

130

Scheme 3.49. Cationic heterobimetallic complex **130** bearing a gold(I) and ruthenium(II) metal center.

As the ruthenium(II) is known to be not stable for extended periods of time,[102] it was planned to insert the gold(I) first, followed by the ruthenium(II). The overall reaction scheme starts from the parent [2.2]paracyclophane that is dibrominated to give 4,16-dibromo[2.2]paracyclophane (**30**). A selective monolithiation and electrophilic trapping with *N,N*-dimethylformamide (DMF) yields 4-bromo-16-formyl[2.2]paracyclophane

(**45**). The direct oxidative conversion of aldehyde **45** to oxazoline **131** was achieved using 1,2-aminoethanol and NBS. Then, a one-pot reaction comprised of lithiation, trapping with Chlorodiphenylphosphine to get the free phosphine and *in situ* auration with tetrahydrothiophene gold(I) chloride (**107**) afforded the gold(I) complex **132**. This complex was cyclometallated in the next step with the dimeric ruthenium precursor $[RuCl(p\text{-cymene})]_2$, finally giving heterobimetallic complex **130** in an overall yield of 39% over five steps (Scheme 3.50).

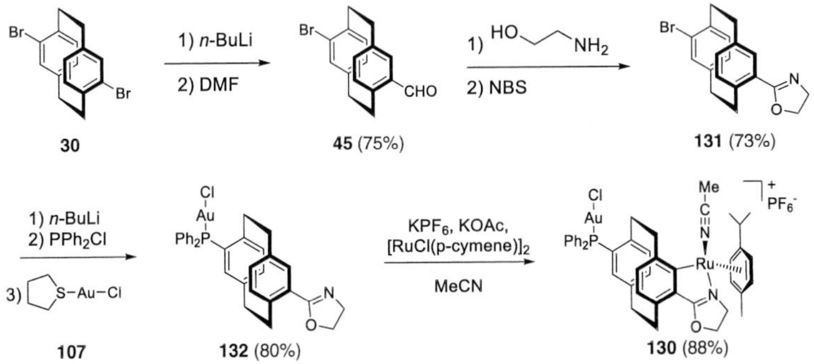

Scheme 3.50. Full synthetic route to prepare heterobimetallic complex **130** in 5 steps with an overall yield of 39%.

Unfortunately, this complex is prone to decomposition and turns from a canary yellow to mud-green in a matter of days even under argon at –20 °C, which has been reported for ruthenium complexes of this sort.[102] Therefore, no suitable single crystals of **130** for X-ray diffractometry could be obtained. However, the molecule could be characterized by means of NMR and HRMS. The ^{31}P NMR shows a singlet at $\delta = 35.3$ ppm which is significantly shifted downfield in comparison to other compounds containing a PCP-P–Au moiety.[157] Additionally, a heptet with a chemical shift of $\delta = -139.2$ ppm is also present in the ^{31}P NMR, corresponding to the PF_6-Anion. The simulated HRMS pattern fits the experimental data very well (Figure 3.15).

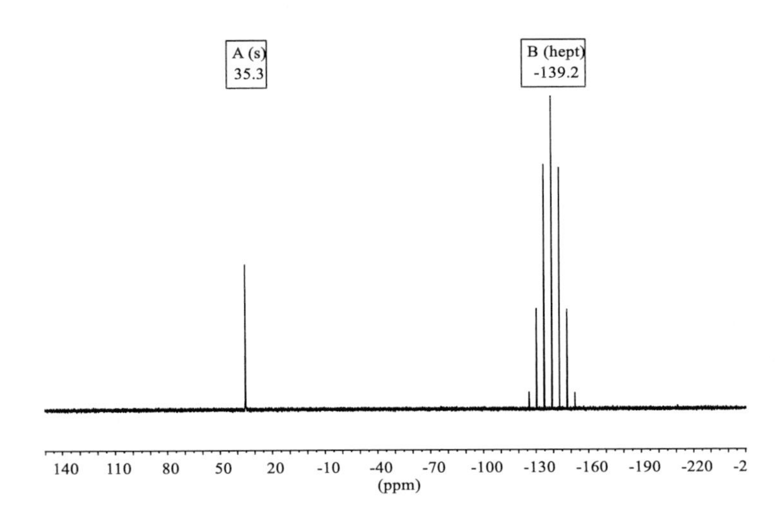

Figure 3.14. ^{31}P NMR spectrum of heterobimetallic complex **130**.

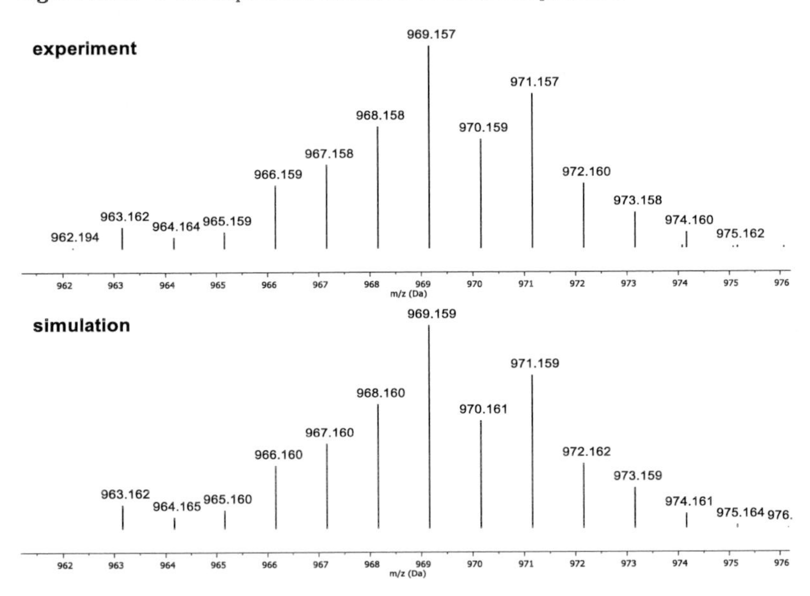

Figure 3.15. Experimental vs simulated ESI MS for complex **130.**

After this first successful synthesis of a heterobimetallic PCP complex, the next goal was to obtain a catalytically active ruthenium center. The η^6-bound *p*-cymene ligand in **130** is bound too tightly to the ruthenium core to allow for the generation of a vacant coordination site, and thus not suited for catalysis. This was planned to be circumvented by preparation of the more labile cyclometallated tetraacetonitrile complex **134** (Scheme 3.51). However, after an initial test synthesis of the ruthenated PCP **133**, it became apparent that the stability of this complex is inferior to the *p*-cymene complex **130**, and decomposes in a few seconds on air. Thus, isolation of octaedric tetraacetonitrile coordinated ruthenium(II) complexes was not further pursued.

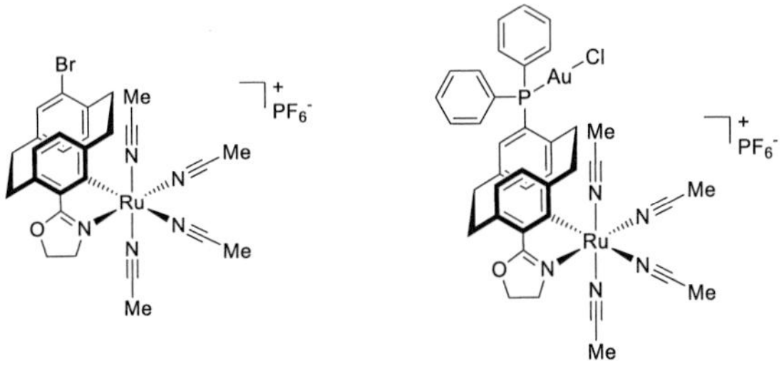

133 **134**

Scheme 3.51. Cationic heterobimetallic complexes **133** and **134**. Complex **133** decomposed in a matter of seconds, dissuading from further attempts to synthesize **134**.

While trying to grow suitable X-ray crystals of the various compounds involved in these syntheses, crystals formed from gold(I) complex **132**. Upon analysis, these crystals turned out to be a decomposition product of **132** and their identity was determined to be the acid-catalyzed ring-opened product **135** (Scheme 3.52, Figure 3.16). Ultimately, this result confirms the pseudo-*para*-configuration proposed for this synthesis route.

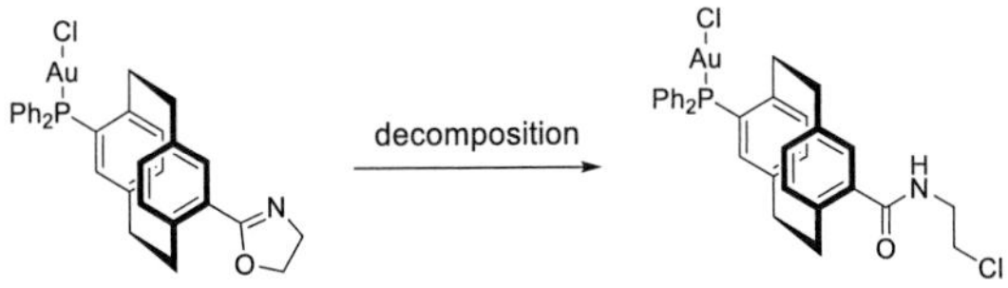

decomposition

132 **135**

Scheme 3.52. Decomposition of gold(I) complex **132** to the ring-opened β-chloroamide **135**.

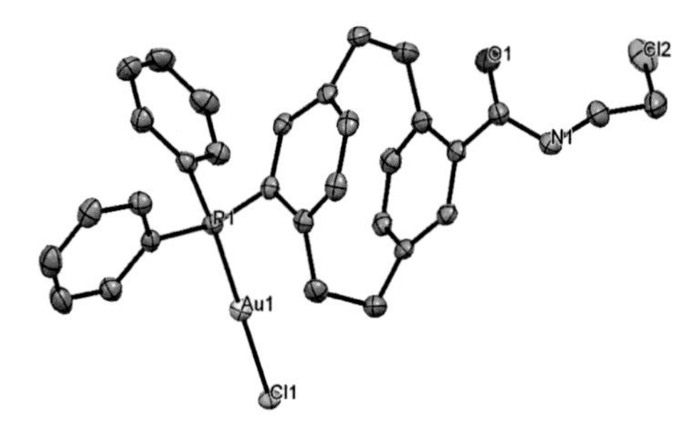

Figure 3.16. Molecular structure of **135,** the decomposition product of **132**. Hydrogen atoms have been omitted for clarity.

3.5.4 Heterobimetallic Complexes of Gold and Palladium

Another path towards a heterobimetallic complex bearing two catalytically active metal centers was sought after. PCP oxazolines are not only suitable for cycloruthenation, but also for cyclopalladation. These PCP C–N palladacycles have been shown to be active in Suzuki cross-coupling reactions.[158] The general synthesis of this class of complexes from oxazoline **129** is accessed through dimer **136** that can be cleaved with triphenylphosphine to afford the desired monomeric structure **137** (Scheme 3.53).

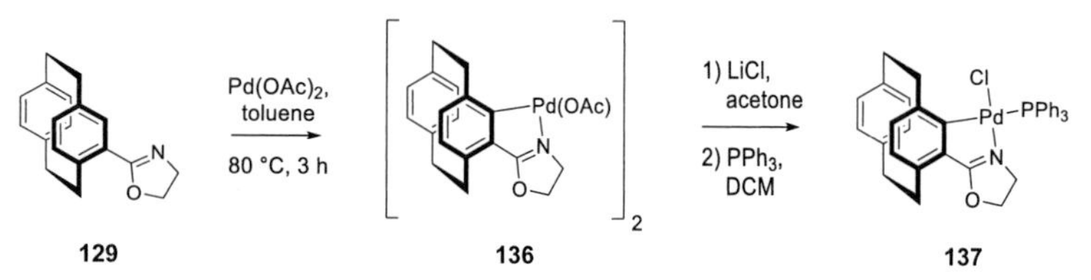

Scheme 3.53. Palladation of oxazoline **129** leads to dimer **136** which can be cleaved with triphenylphosphine to afford monomeric **137**.

Interestingly, the solvent used in the cyclopalladation step determines the regioselective outcome of this reaction. If acetic acid is used instead of toluene, the cyclopalladation occurs at the bridge carbon (Scheme 3.54).[74]

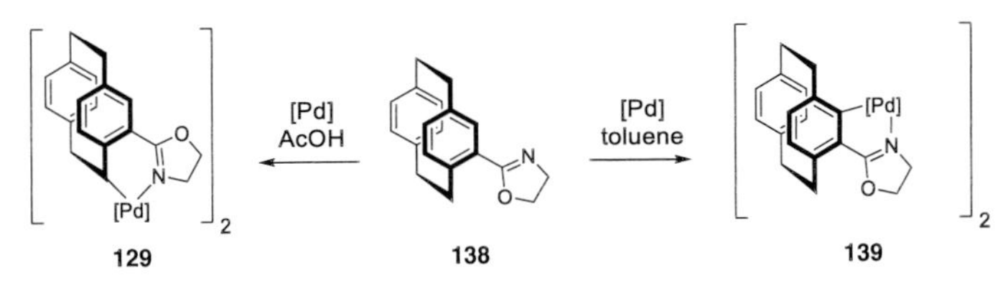

Scheme 3.54. Regioselective palladation of PCP oxazoline **129** depending on the solvent used.

Accordingly, the cyclopalladation reaction was run with the gold(I) complex **132** to access heterobimetallic gold-palladium complex **140** (Scheme 3.55). Even before heating the reaction mixture, the solution turned black upon the addition of palladium(II) acetate. A ^{31}P NMR of the crude reaction mixture revealed more than 7 different phosphine species, none of them corresponding to the starting material or even being in the range of expected P–Au peaks. A black precipitate could be observed which could stem from precipitated palladium(0).

Cl
Au
Ph_2P
$Pd(OAc)_2$,
toluene
80 °C, 3 h
black precipitate
Pd(OAc)
132
140

Scheme 3.55. Unsuccessful cyclopalladation of gold(I) complex **132**. Instead, a black precipitate was formed.

The inverted route, first introducing palladium, then gold, was tried next. Since palladium is also easily coordinated by free phosphine coordination sites, it was planned to use the borane-protected phosphine **141** (Scheme 3.56). However, upon addition of the palladium(II) acetate, the solution turned black once more and gas evolution was observed. A literature research revealed that indeed palladium(0) particles are suited for dehydrogenation of amine-boranes which are very similar in their reactivity to phosphine-boranes, supporting the hypothesis of reduction of Pd(II) to Pd(0).[159–161]

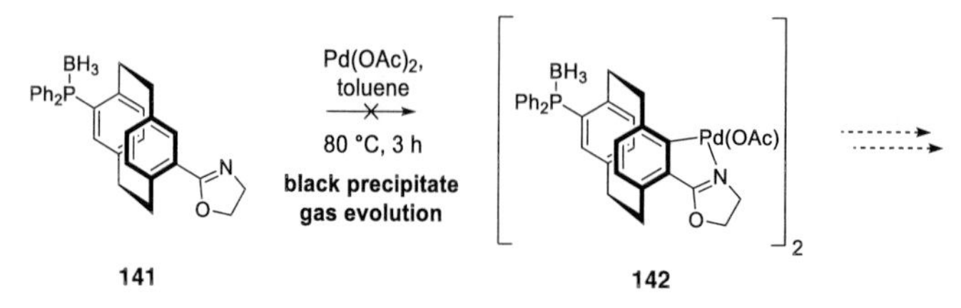

Scheme 3.56. Unsuccessful cyclopalladation of phosphine borane **141**. Instead, a black precipitate formed, and gas evolution was observable.

3.5.5 Conclusions

It was aimed to prepare proof-of-principle target complex **130** (Scheme 3.57) according to the strategy presented in chapter 3.1. It was shown that the synthesis of heterobimetallic gold-ruthenium complexes is possible by the preparation of a ditopic ligand and stepwise metalation with suitable precursors. The relative configuration of the two substituents on the PCP core was confirmed by X-ray crystal structure analysis of a decomposition product since no other attempts were successful. The stability of the ruthenium(II) *p*-cymene complexes, while beneficial for purification and analysis, prevents their application in catalysis. On the other hand, the more labile octaedric tetraacetonitrile coordinated ruthenium(II) complexes were not stable enough for storage or catalysis purposes. Therefore, a different coordination environment for ruthenium is probed in chapter 3.6. The preparation of gold-palladium heterobimetallic complexes remains challenging, as only decomposition was observed after the addition of palladium(II) acetate in the final step. Thus, this path was not further pursued.

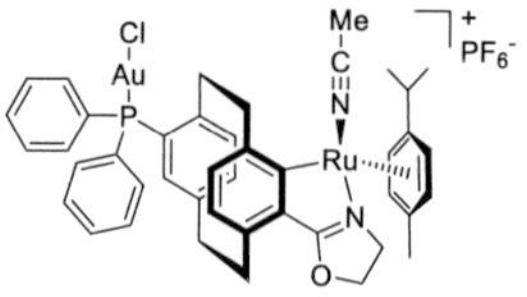

130

Scheme 3.57. The successfully prepared ruthenium(II)-gold(I) complex **130** confirmed the viability of the strategy used to obtain heterobimetallic complexes.

3.6 Photoredox Catalysis with PCP Au/Ru Complexes

3.6.1 Introduction

After the first successful hit on heterobimetallic complexes, the gold(I)-phosphine moiety presented in chapter 3.4 was selected for one of the two metal centers. For the other metal center, the strategy was changed to turn towards a more stable coordination environment for the ruthenium(II) in the planned heterobimetallic complexes.

One of the most stable ruthenium(II) complexes is the widely used photocatalyst $Ru(bpy)_3Cl_2$ or the corresponding hexafluorophosphate salt $[Ru(bpy)_3]PF_6$ **143** (Scheme 3.58). This photocatalyst has found ample application in photoredox catalysis.[162]

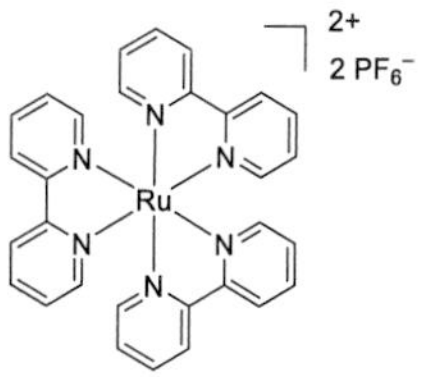

143

Scheme 3.58. The widely used photocatalyst **143**.

Recently, reports about variants of **143** were published that replaced one of the 2,2'-bipyridine substituents with a phenylpyridine, turning the complex into the monocationic cyclometallated ruthenium photocatalyst **144**.

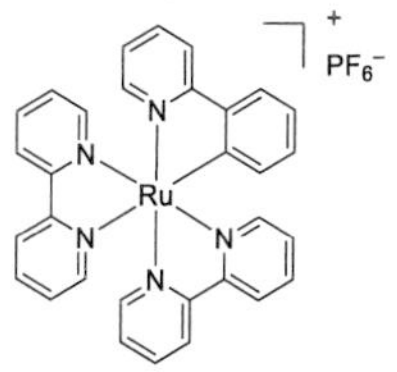

144

Scheme 3.59. Cycloruthenated variant of photocatalyst **143**, the monocationic ruthenium(II) complex **144**.

While showing similar stability, cycloruthenated **144** shows a much wider electronic absorption spectrum, thus allowing for photons of wavelengths of up to 700 nm to excite the complex. This promises a much better capability as photocatalyst compared to the relatively early cut-off at around 550 nm of **145** (Figure 3.17).[163]

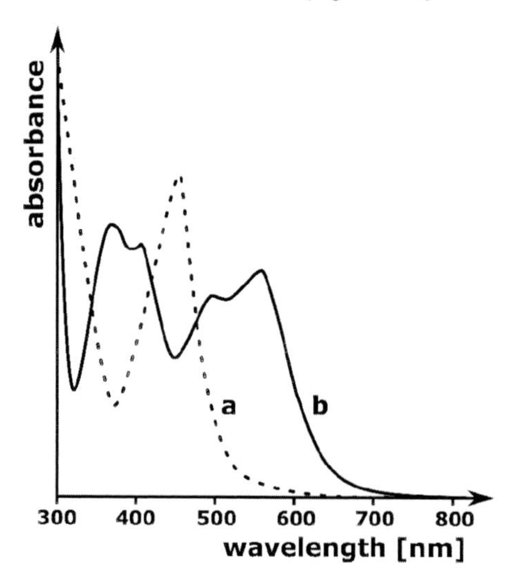

Figure 3.17. Electronic absorption spectrum of **143** (**a**) and **144** (**b**). The cycloruthenated **b** shows a much wider absorption window.[163]

Above considerations led to the decision to aim for pseudo-*para* heterobimetallic PCP complex **145** and its *ortho* and *geminal* isomers (Scheme 3.60).

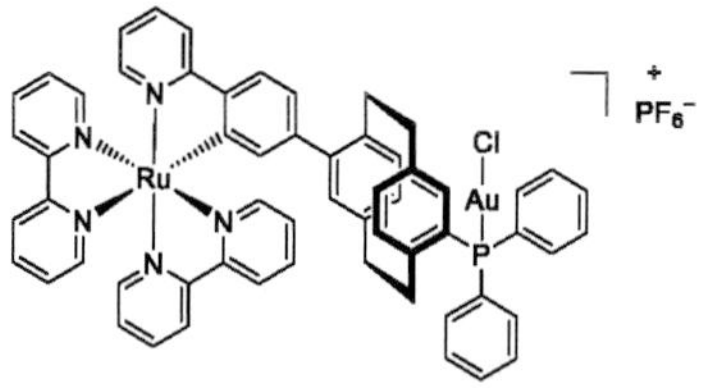

145

Scheme 3.60. Target pseudo-*para* heterobimetallic gold-ruthenium complex **145**.

These isomers were thought to be the ideal model platform to probe the distance dependency of gold(I) catalyst and ruthenium(II) photocatalyst in a cooperative visible-light driven photoredox catalysis.

3.6.2 Synthetic Access to the pseudo-*para* Au/Ru Heterobimetallic Complex

A closer analysis of target complex **145** shows that, starting from commercially available 4,16-dibromo[2.2]paracyclophane (**30**), a few synthetic considerations arise: (i) The heterodisubstitution can be achieved easily with the methods presented in chapter 3.1. (ii) The phosphine can be made with methods presented in chapter 3.5. (iii) The free phosphine (prior to auration) is prone to oxidation and either must be protected (oxide/sulfide/borane) or aurated before other synthetic steps or purification by column chromatography can be done. (iv) The free phosphine would coordinate to ruthenium(II) in the cyclometallation step, thus leading to unwanted side products. Thus, this route should be avoided. (v) While stable to most conditions, the bipyridyls are electron-deficient to begin with which is pronounced after coordination. Attack by nucleophiles could become a problem. (vi) The common use of silver salts to do the cycloruthenation step could cause issues with the pre-installed gold(I) complex. Thus, it is advisable to install the ruthenium first.

3.6.2.1 Replacing bromine with/without Ru installed

In section 3.5.3, a one-pot protocol comprised of metal-halogen exchange followed by trapping with chlorodiphenylphosphine and *in situ* auration was used to transform an aryl bromide to the respective phosphine gold(I) complex (Scheme 3.61). Although problems regarding a nucleophilic attack at the bipyridyls might arise, the prior successfully implemented method was probed as well.

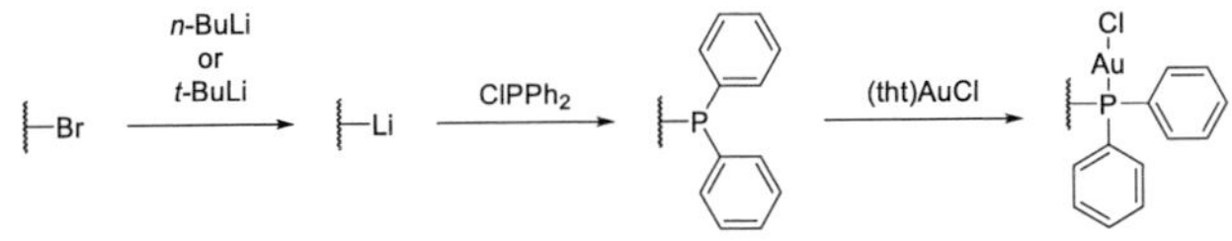

Scheme 3.61. One-pot *in-situ* metal-halogen exchange, phosphination and auration to transform an aryl bromide into the respective phosphine gold(I) complex.

As well, this was applied to introduce phosphine-gold(I) in the ruthenium(II) complex **146** (Table 6). To ensure complete conversion of the starting material, a three-fold excess of all reagents was used (Table 6, entry 1). This was especially necessary because the small amounts of chlorodiphenylphosphine. While the product was detected by ESI MS, reaction control by TLC revealed a mixture of side products. Attempts to isolate the desired heterobimetallic complex **145** by column chromatography on silica failed. Instead, the crude mixture contained a substantial amount of the decomplexed and debromination product **147**. Debromination can be explained by successful lithiation of the aryl bromide and subsequent quenching by either not completely dry solvents or impure chlorodiphenylphosphine, which can contain significant amounts of diphenylphosphinic acid if exposed to air. This phosphinic acid is an excellent source of protons, capable of quenching the lithiated PCP. In addition, alkylated product **148** was detected by ESI MS in the crude reaction mixture. The exact location of the alkyl chain could not be elucidated, but alkylation of bipyridines with alkyllithium reagents is described in literature.[164]

Table 6. Unsuccessful one-pot construction of phosphine-gold(I) part of heterobimetallic complex **145**.

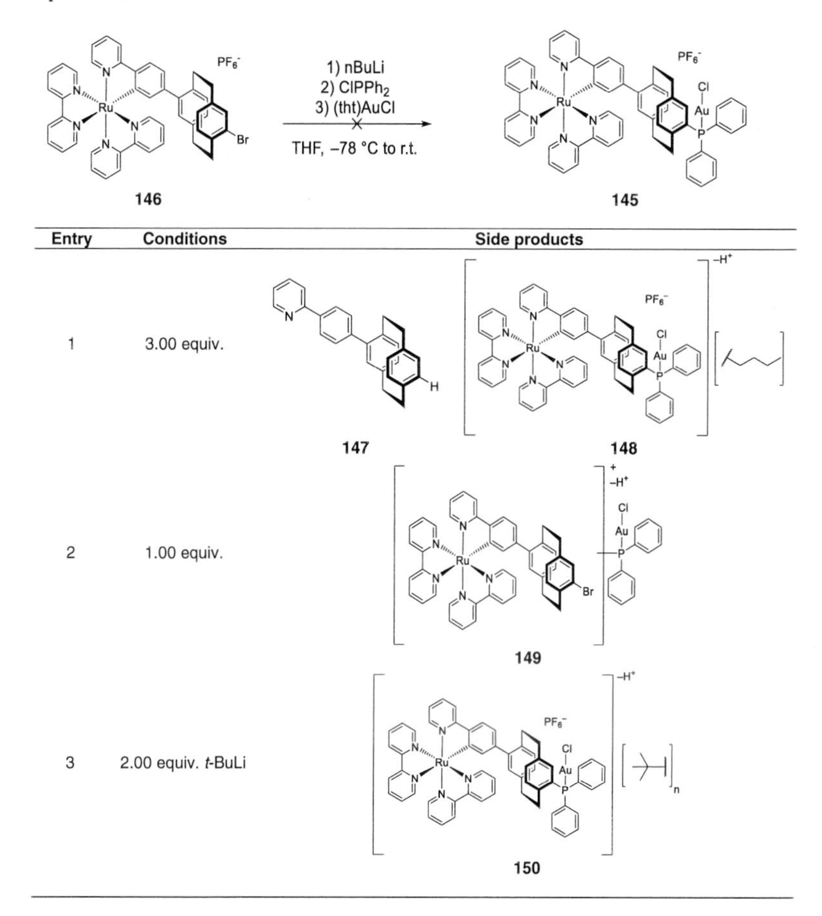

Entry	Conditions	Side products
1	3.00 equiv.	147, 148
2	1.00 equiv.	149
3	2.00 equiv. *t*-BuLi	150

The reaction was repeated on a larger scale with 1.00 equiv. of all reagents to avoid alkylation side products. However, ESI MS revealed that the main peak corresponds to $[M–H+Br]^+$, with M being heterobimetallic complex **145**. A possible explanation for this could be a deprotonation of an aryl C–H by the organolithium followed by quenching of this carbanion by chlorodiphenylphosphine and subsequent auration.

Next, replacement of *n*-butyllithium by the less nucleophilic *tert*-butyllithium was tested. However, analysis of the crude reaction mixture by ESI MS indicated a complex mixture containing species **150** with different degrees of alkylation caused by *tert*-butyllithium.

As the dicationic ruthenium significantly decreases the electron density of the phenylpyridine and bipyridine ligands, a nucleophilic attack of organolithium reagents on these ligands becomes more facile. Therefore, it was tested next if the free phenylpyridine ligand is more suitable to successfully transform the remaining bromide to the gold(I) complex by the same methodology.

Subjecting the free ligand **151** to the same conditions as above likewise led to alkylated debromination product **153** and alkylated gold(I) complex **154** (Table 7). Since Grignard reagents are less nucleophilic and less basic than their organolithium counterparts,[165] a turbo-Grignard reaction in the first step was tested to achieve the desired one-pot transformation without alkylation side products. Analysis of the crude reaction mixture by ESI MS shows the product **152** in small amounts accompanied by a range of side products: the product gold(I) complex with bromide instead of chloride **155**, debromination product **147**, traces of phosphine oxide **156**, and unreacted starting material **151**. The formation of **155** can be explained by the known halide exchange of chloride in gold(I) complexes in the presence of bromide.[166] If the reaction mixture is not entirely degassed or the auration step is not complete when the reaction is quenched and exposed to air, phosphine oxide **156** is formed from the intermediate free phosphine.

Table 7. Unsuccessful one-pot construction of phosphine-gold(I) part of monometallic complex **145**.

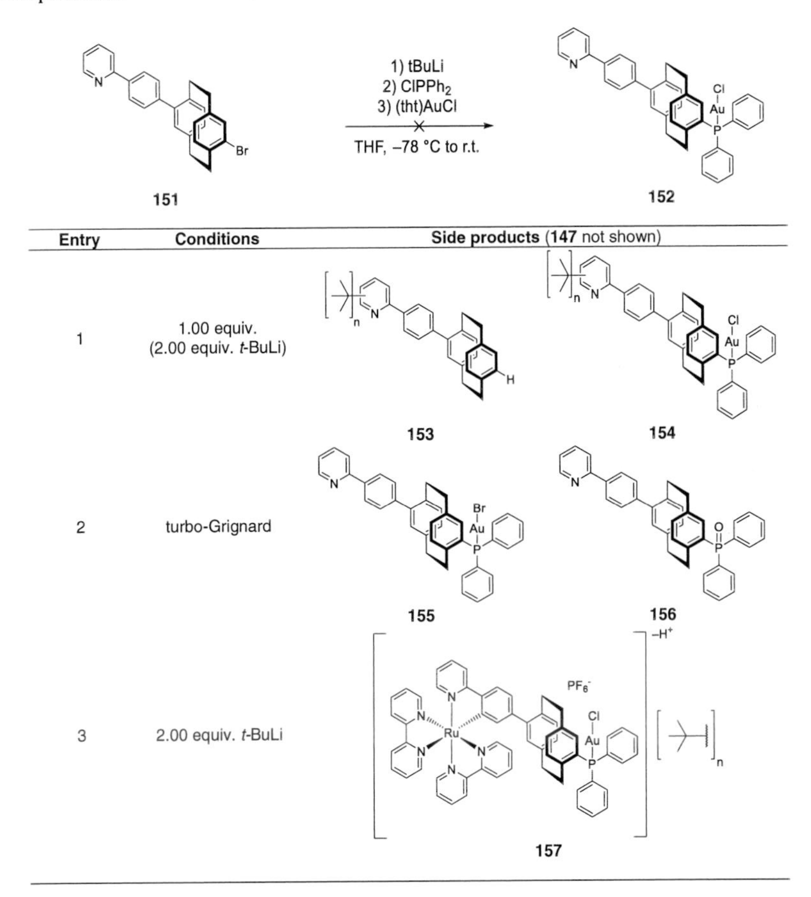

Entry	Conditions	Side products (147 not shown)
1	1.00 equiv. (2.00 equiv. *t*-BuLi)	153, 154
2	turbo-Grignard	155, 156
3	2.00 equiv. *t*-BuLi	157

3.6.2.2 Cross-coupling as an alternative strategy

To circumvent the use of alkyllithium reagents, a copper-[167] and a palladium-catalyzed[168] direct C–P cross-coupling method were tested on substrate **151** (Scheme 3.62). However, ESI and ASAP MS revealed only starting material and debromination product **147** for both reactions.

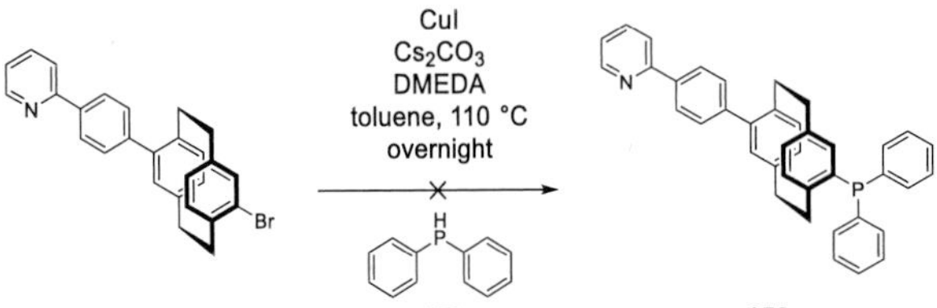

151 158 159

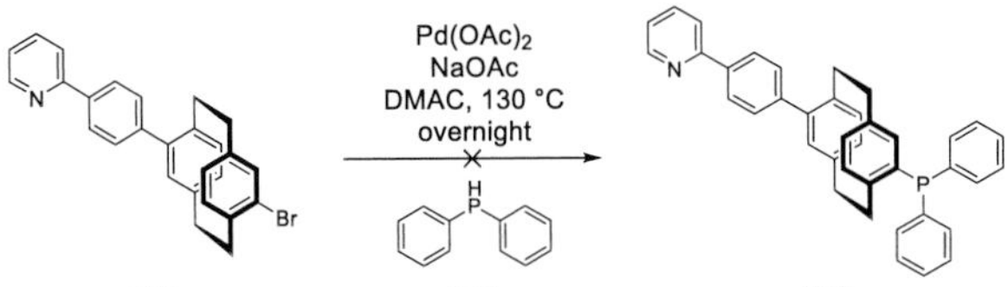

151 160 161

Scheme 3.62. Unsuccessful direct C–P cross-couplings of free ligand **151** with diphenylphosphine (**158**).

3.6.2.3 Construction of the free phosphine and C–N ligand

After the unsuccessful route employing phenylpyridine substituted PCP **96**, the focus was changed to the inverse route. Installing the phosphine, oxidizing it on purpose, cross-coupling with phenylpyridine and subsequent reduction of the phosphine should lead to the free phosphine.

For the first step, the methodology presented in section 3.2.1.2 can be applied (Scheme 3.63).

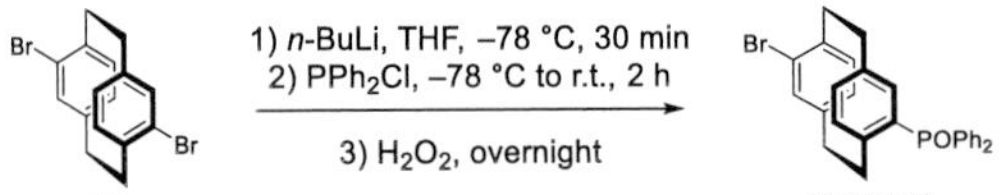

30 **44** (96%)

Scheme 3.63. Synthesis of phosphine oxide **44** by selective monolithiation and electrophilic trapping, followed by oxidation.

This reaction is very high yielding and can be run on a multi-gram scale. Next, the introduction of phenylpyridine was performed by Suzuki cross-coupling using the optimized conditions for PCP presented in section 3.3. The reaction proceeds smoothly to afford **163** in 93% yield. TLC control of the reaction revealed a color change from colorless (all starting materials) to blue fluorescing (**163**) under 265 nm UV light.

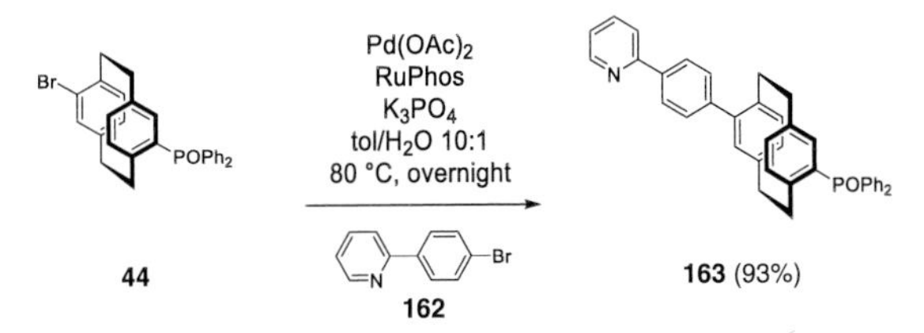

Scheme 3.64. Synthesis of phenylpyridine substituted PCP **163**.

With the product **163** in hand, the next step is to reduce the phosphine oxide. Several methods are reported in literature,[169-173] whereas the conditions which are used to prepare Phanephos were found to be the most fit.[24] Using trichlorosilane in a high boiling unpolar solvent, the reduction of **163** is achieved with the conditions shown in Scheme 3.65.

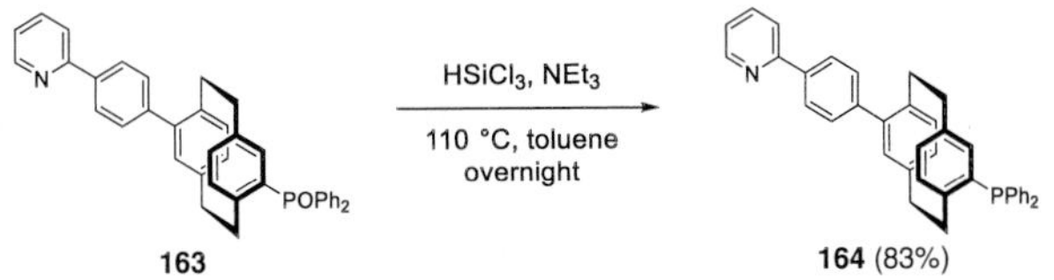

Scheme 3.65. Reduction of phosphine oxide **163** to the free phosphine **164** with trichlorosilane at high temperatures.

Since the free phosphine is prone to oxidation, purification by chromatography is not possible without significantly reduced yield. Therefore, the crude mixture was subjected to a fast aqueous workup, analyzed by ^{31}P NMR and used in the next step without further purification. The peak at $\delta = 27.2$ ppm corresponds to the phosphine oxide 163, while $\delta = -3.5$ ppm corresponds to the free phosphine **164** in the ^{31}P NMR (Figure 3.18). Both of these chemical shifts are in good agreement with values reported in literature.[157,171] The conversion was determined to be 83%.

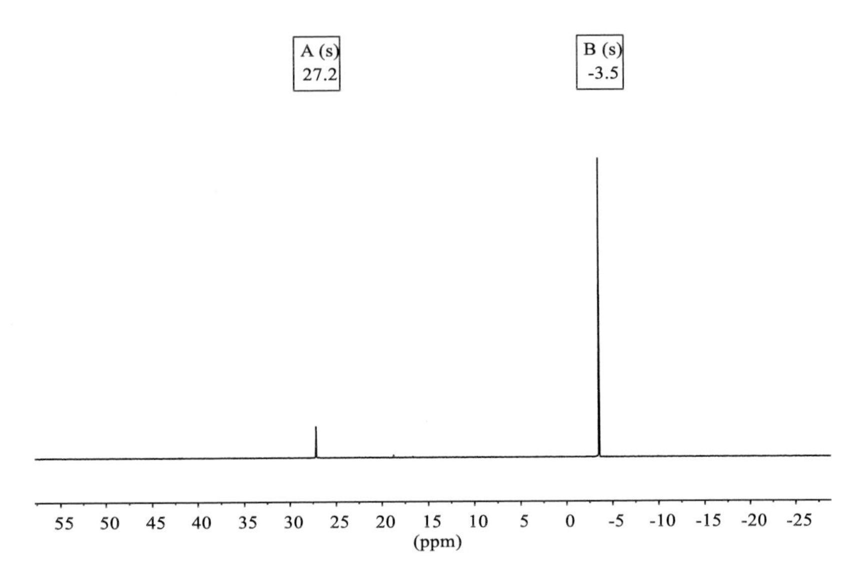

Figure 3.18. Crude ^{31}P NMR of the phosphine oxide **163** (A) reduction to the free phosphine **164** (B).

The next reaction step was done with the crude mixture of the reduction reaction (Scheme 3.65). The crude phosphine was reacted with (tht)AuCl. The labile tetrahydrothiophene ligand is easily replaced by the strongly bonding phosphine to afford gold(I) complex **165** in 46% yield (Scheme 3.66).

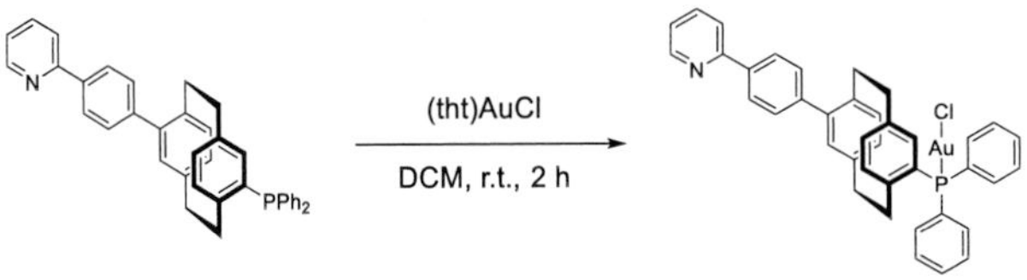

(tht)AuCl

DCM, r.t., 2 h

164 **165** (46%)

Scheme 3.66. Coordination of phosphine ligand **164** to (tht)AuCl to afford gold(I) complex **165**.

The reaction could be monitored by means of ^{31}P NMR that shows a peak at $\delta = 29.8$ ppm, an expected significant downfield shift from the free phosphine to the phosphine gold(I) complex. The chemical shift corresponds well to values reported in literature for similar compounds.[157]

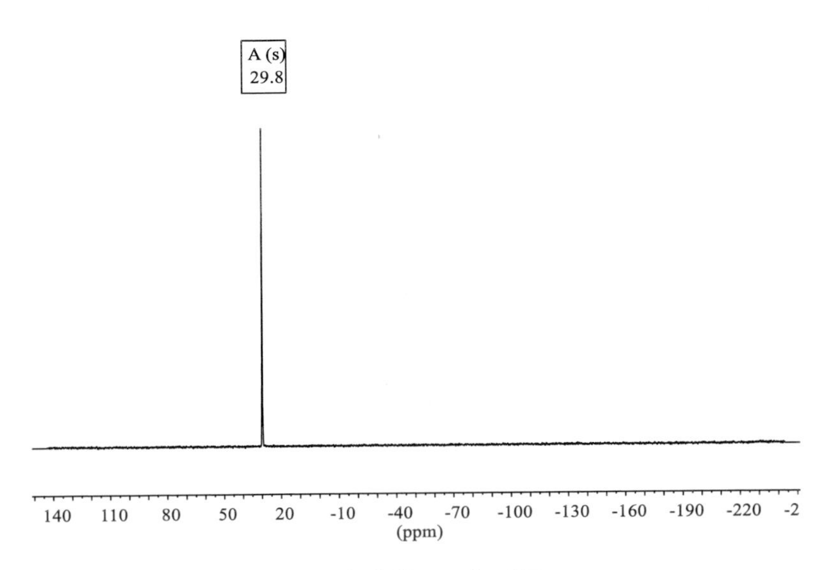

Figure 3.19. ^{31}P NMR spectrum of gold(I) complex **165**.

The structure of the complex was further analyzed by collision induced dissociation (CID) ESI MS. The base spectrum was recorded at collision energy E = 0 eV. The peak of highest intensity at m/z = 778.2 Da corresponds to the protonated compound **165**. When the collision energy was raised to 60 eV, **165** dissociates and the two fragments **166** and **167** could be detected. Fragment **166** is formed through heterolytic cleavage of the Au–Cl bond. Fragment **167** is formed through a typical cleavage of PCPs that involves the separation of the two decks analogous to the separation of the two halves of a sandwich. This fragment further serves as proof of the structure and shows that the substituents indeed are not located on the same deck.

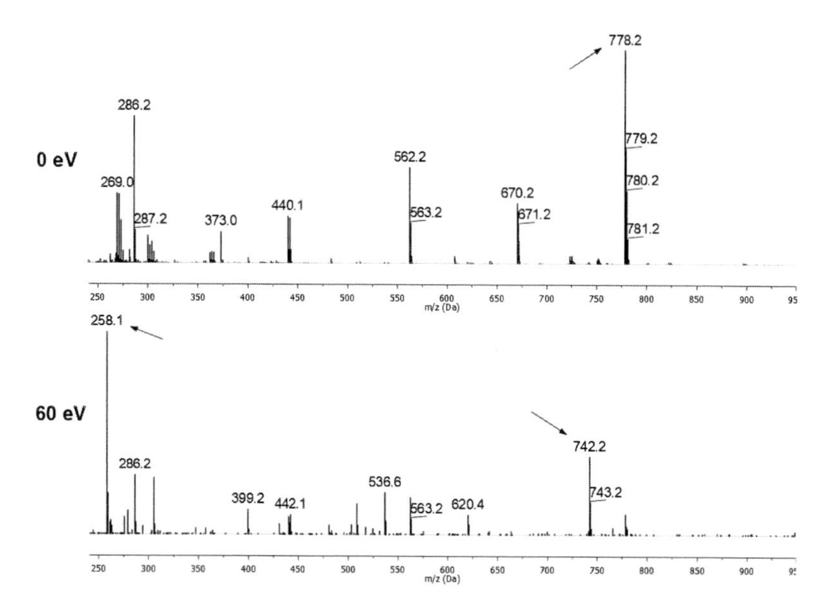

Figure 3.20 CID ESI MS of gold(I) complex **165** at 0 and 60 eV.

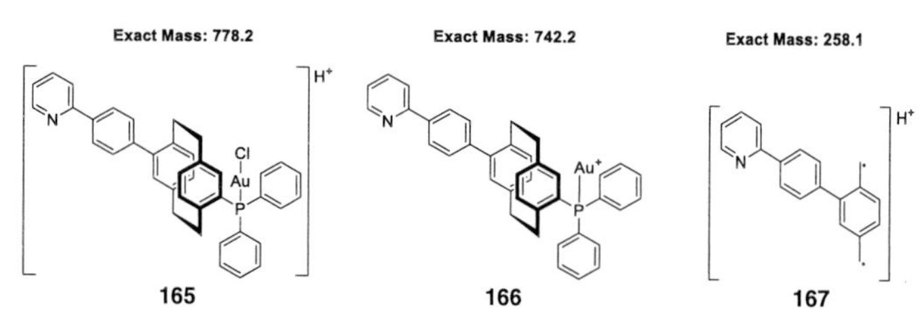

Figure 3.21. The corresponding fragments found by CID ESI MS of gold(I) complex **165**.

The molecular structure of **165** was also confirmed by X-ray single crystal diffractometry (Figure 3.22). The phenylpyridine moiety (plane through C17-C22) is twisted out of the PCP deck plane (C12-C16) by 41.8°. This angle is significantly larger than comparable PCP pyridyl scaffolds.[88] The P–Au and Au–Cl distances are in the expected range with bond lengths of 2.236 Å and 2.292 Å respectively.[157] The gold atom is almost linearly coordinated, displaying a gold(I) typical twofold coordination mode [P–Au–Cl: 176.2°]. The P-Au-Cl "tail" is turned towards the PCP scaffold rather than the phenyl substituents to form a 98.6° torsion angle (C5–C4, P–Au).

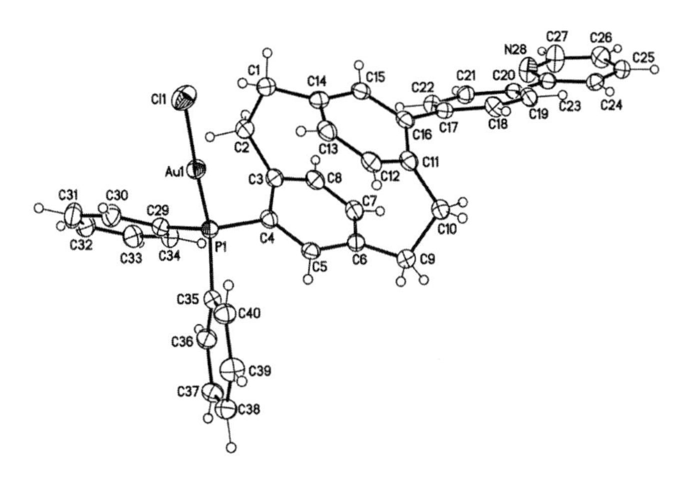

Figure 3.22. Molecular structure of **165/SB1259_sq** (displacement parameters are drawn at 50% probability level).

3.6.2.4 Ruthenation of gold(I) complex **165**

The final step in the synthesis of heterobimetallic target complex **145** is the cycloruthenation with a suitable ruthenium precursor. The general preparation of $[Ru(bpy)_2(ppy)]PF_6$ (**169**) is done by treatment of Hppy ligand **16** and $Ru(bpy)_2Cl_2 \cdot 2\,H_2O$ (**168**)[174] with ammonium hexafluorophosphate to provide the PF_6 anion and silver tetrafluoroborate to deprotonate Hppy, precipitate AgCl and in turn generate free coordination sites at the ruthenium center (Scheme 3.67).[175] These conditions were tested for the preparation of gold-ruthenium complex **169** (Scheme 3.68). Although the analysis by ESI MS revealed the presence of complex **145**, only traces could be obtained after column chromatography.

H_2O H_2O ; $AgBF_4$, NH_4PF_6 ; DCM, reflux, 2 h ; PF_6^-

16 + **168** → **169** (70%)

Scheme 3.67. General preparation of cycloruthenated $[Ru(bpy)_2(ppy)]PF_6$ (**169**).[175]

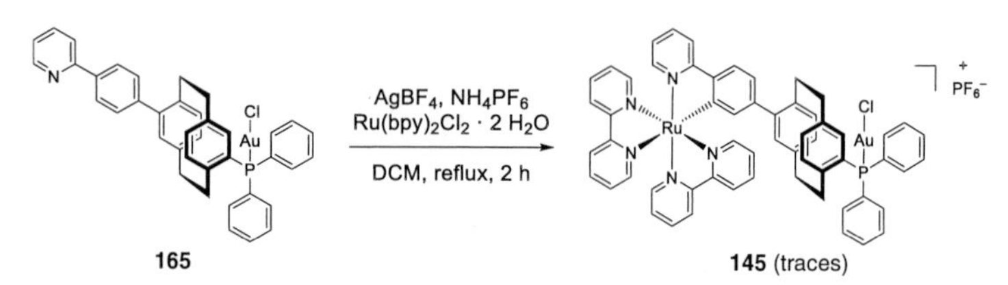

Scheme 3.68. Attempted cycloruthenation that generated only traces of heterobimetallic complex **145**.

The solution turned from a deep purple to nearly colorless when the mixture was brought to a reflux. This disappearance of color indicates the decomplexation of ruthenium from any bipyridine or phenylpyridine ligand, as both the starting material ruthenium complex **168** and the product **145** are intensely colored. Since silver salts are commonly used to transform neutral gold(I) chlorides into cationic gold complexes,[176] it is likely that the addition of silver in this reaction not only abstracted the chloride from the ruthenium precursor as planned, but also broke the Au–Cl bond, leaving the reaction with a cationic gold(I) complex. It is postulated that this reactive intermediate reacts with the ruthenium species in solution, thus destroying the ruthenium(II) complex. Therefore, it was necessary to avoid the silver salt route in order to prepare the bimetallic complex **145** starting from the gold(I) chloride complex **165**.

The first test reaction without using the silver salt route was done by replacing $AgBF_4$ with KOH serving as a base to deprotonate the ligand **165** and thus facilitating cyclometallation. The mixture retained the purple color even at 80 °C, which indicates that the ruthenium was still in its complexed form. However, only traces of the product could be isolated. The same side product as in the silver salt run was detected, along with a new compound that shows a mass spectrum matching structure **170** (Figure 3.23).

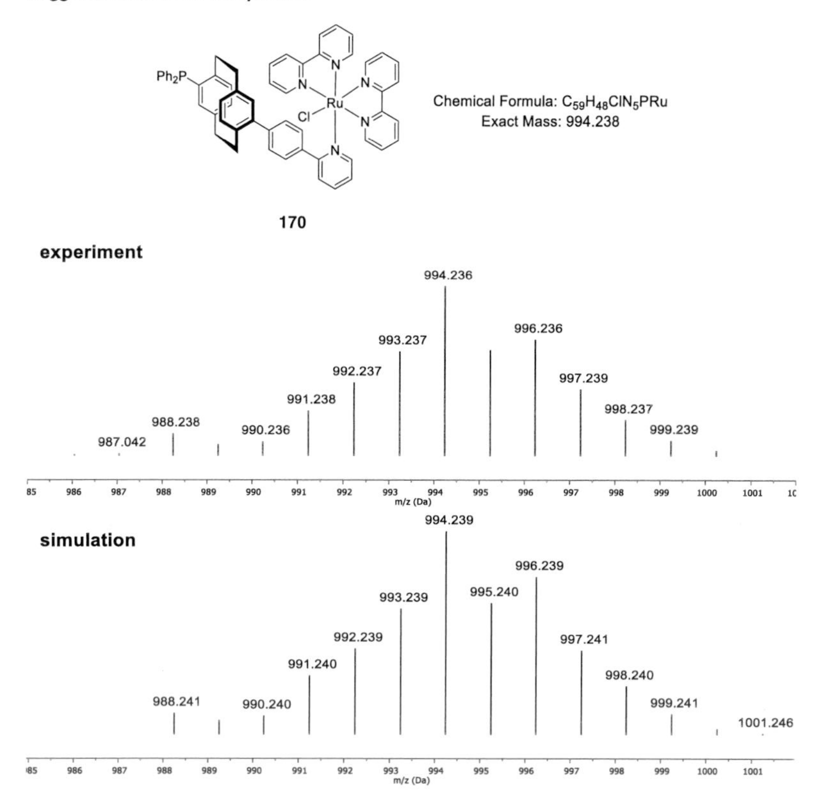

Figure 3.23. Plausible structure suggestion for side product **170** with experimental vs. simulated positive mode ESI mass spectrum.

It is hypothesized that the gold(I) chloride can aid in partially abstracting the chloride off of $Ru(bpy)_2Cl_2 \cdot 2\ H_2O$ (**168**), thereby being removed from the phosphine coordination site and affording complex **170**. Unfortunately, the amount of material obtained by column chromatography was too small to record a ^{31}P NMR to evaluate the chemical environment of the phosphorous atom in **170,** thus the structure remains hypothetical.

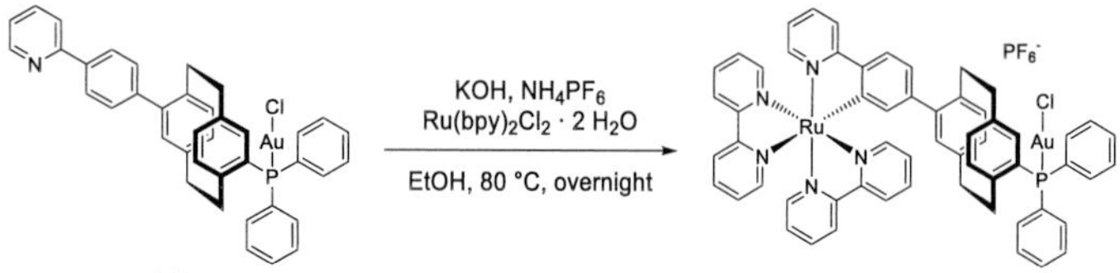

165 145 (traces)

Scheme 3.69. Attempted cycloruthenation of **165**. KOH used as base to avoid the silver salt route.

These findings hinted toward the chloride abstraction being the key step in this complexation reaction and thus it was planned to circumvent the chloride abstraction by replacing them with more labile pyridine ligands. To this end, the synthesis route was changed slightly to be more convergent and performing the chloride abstraction in a separate reaction step on $Ru(bpy)_2Cl_2 \cdot 2\ H_2O$ (**168**). Treatment of **168** with a large excess of pyridine in refluxing water affords $[Ru(bpy)_2(py)_2]Cl_2$ (**171**) in excellent yield (Scheme 3.70). This chloride salt is reported to undergo efficient ligand exchange with phenylpyridine.[177] However, analysis of the reaction conducted analogous to the reported conditions showed only unreacted starting materials (Scheme 3.71).

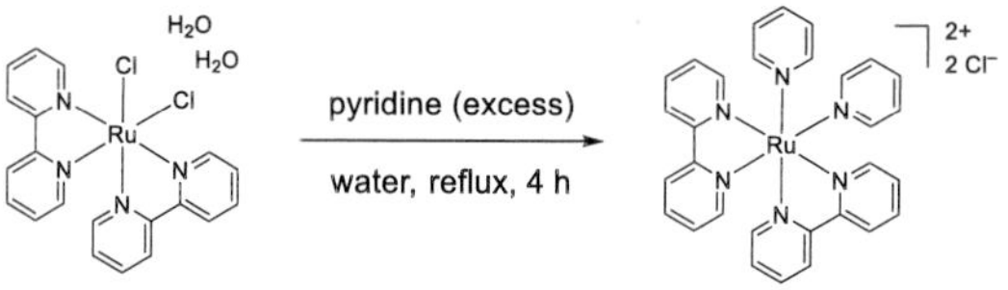

168 171 (94%)

Scheme 3.70. Preparation of chloride salt **171** from precursor **168** by treatment with pyridine and water at reflux.

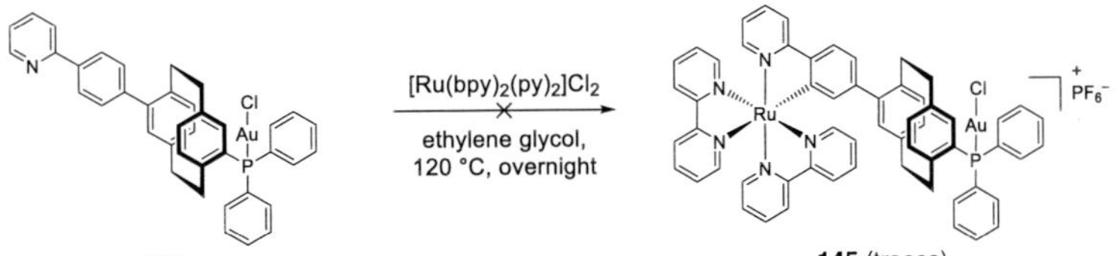

$[Ru(bpy)_2(py)_2]Cl_2$

ethylene glycol, 120 °C, overnight

165 145 (traces)

Scheme 3.71. Unsuccessful cycloruthenation of **165** with $[Ru(bpy)_2(py)_2]Cl_2$ (**171**) to avoid the chloride abstraction in the presence of the gold(I) complex.

Another silver-free cycloruthenation protocol was tested next. Matsui and coworkers were able to use *N*-methylmorpholine as a base to deprotonate C^N ligands with subsequent coordination to ruthenium from the ruthenium precursor **168** (Scheme 3.72).[178] However, applying the same conditions to the PCP gold(I) complex **165** afforded only traces of the desired bimetallic complex 145 (Scheme 3.73).

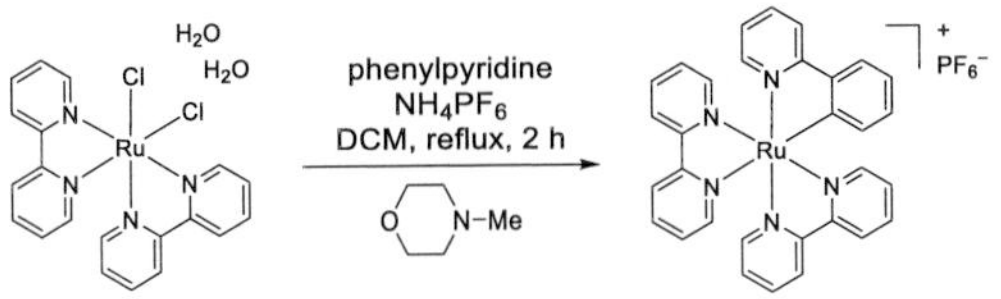

phenylpyridine

NH_4PF_6

DCM, reflux, 2 h

168 172 173

Scheme 3.72. Silver free cycloruthenation employing *N*-methylmorpholine (**172**) as a base.

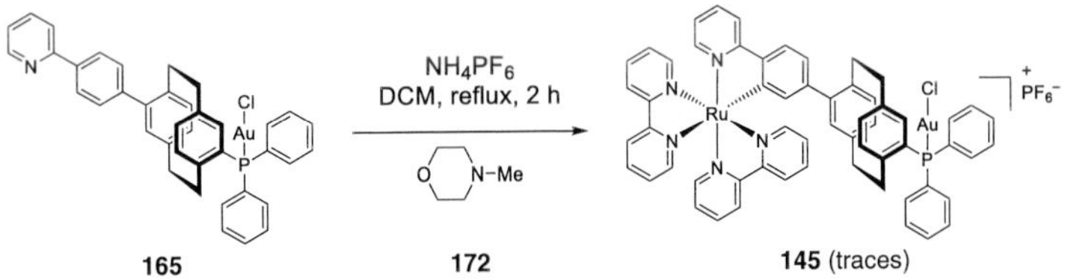

NH_4PF_6

DCM, reflux, 2 h

165 172 145 (traces)

Scheme 3.73. Attempted cycloruthenation of **165** using N-methylmorpholine as a base to avoid the silver method for the abstraction of chloride.

The most promising effort was with KOH as a base (Scheme 3.69). The product **170** that was detected by ESI MS hinted to an incomplete removal of chloride from the ruthenium precursor. Therefore, the reaction was attempted again with 3.00 equivalents of **165**, to make sure enough of gold precursor **165** remains for cycloruthenation after abstraction of the chloride from the ruthenium precursor **168** (Scheme 3.74). However, the outcome of the reaction did not change compared to Scheme 3.69.

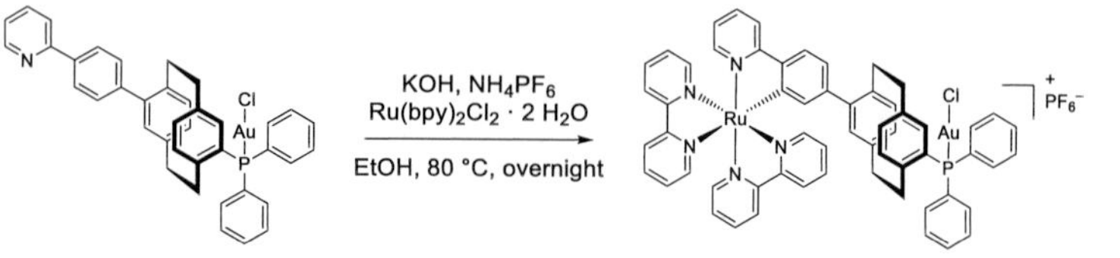

KOH, NH4PF6
Ru(bpy)2Cl2 · 2 H2O
EtOH, 80 °C, overnight
Cl
Au
P
N
Ru
PF6−

165 (3.00 equiv.) **145** (traces)

Scheme 3.74. Attempted cycloruthenation of **165**. A threefold excess of gold complex **165** was used to ensure complete chloride abstraction.

Next, it was tested whether a preforming of the chloride abstracted ruthenium complex could be done *in situ* by refluxing $AgBF_4$ with ruthenium precursor **168** (Scheme 3.75). ESI MS showed that the ruthenium precursor was completely consumed, but no product was formed in this reaction and starting material **165** was still present.

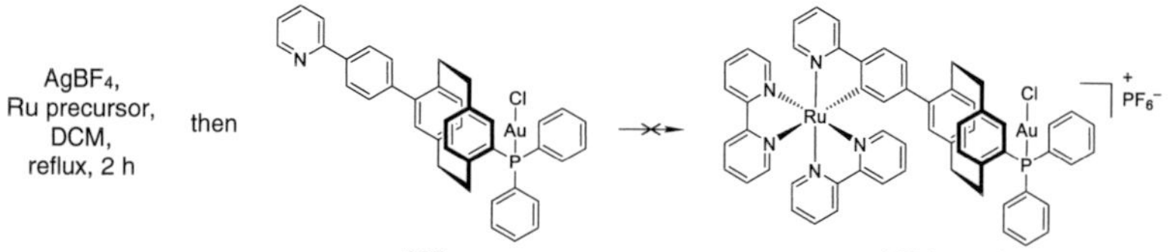

AgBF4,
Ru precursor,
DCM,
reflux, 2 h
then
Cl
Au
P
N
Ru
PF6−

165 **145** (traces)

Scheme 3.75. Attempted cycloruthenation by preforming the chloride abstracted ruthenium complex.

Another method to avoid the silver salt route is to exchange the ruthenium precursor for the $[Ru(benzene)Cl_2]_2$ dimer. This precursor is known to form tetraacetonitrile complexes with cyclometallating C^N ligands in acetonitrile at elevated temperatures.[179] The conditions were applied to the gold complex **165** (Scheme 3.76). The successful formation of the rather labile tetraacetonitrile complex **174** was verified by ESI MS. Complex **174** could be purified by chromatography on silica, where the canary yellow band was isolated from the dark green reaction mixture.

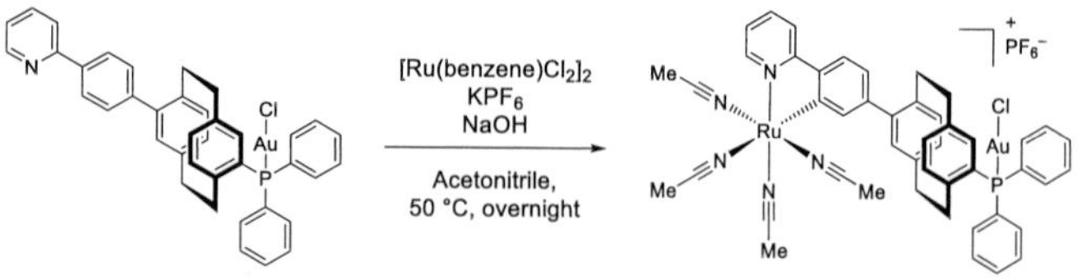

[Ru(benzene)Cl2]2
KPF6
NaOH
Acetonitrile,
50 °C, overnight
Me
N
Ru
Cl
Au
P
PF6−

165 **174**

Scheme 3.76. Cycloruthenation of **165** to afford tetraacetonitrile complex **174**.

The labile acetonitrile ligands can be replaced by chelating bipyridine ligands. Complex **174** was stirred in refluxing methanol with 2,2'-bipyridine for two hours over which the solution turned from canary yellow to an intense dark purple. The desired complex **145** was obtained in very satisfying yield of 78% over two steps from **165** (Scheme 3.77). Characterization by ESI MS and NMR verified the formation of **145**.

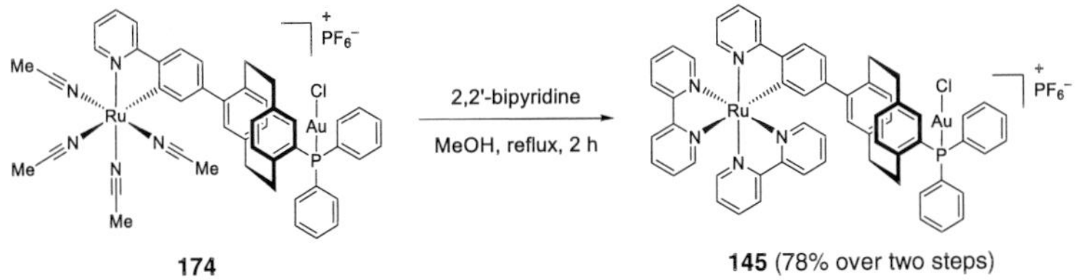

2,2'-bipyridine

MeOH, reflux, 2 h

174

145 (78% over two steps)

Scheme 3.77. Replacement of the acetonitrile ligands in **174** by 2,2'-bipyridine to afford **145** in very satisfying yields.

The ^{31}P NMR (Figure 3.24) of compound **145** shows a heptet at $\delta = -144.2$ ppm that corresponds to the PF_6-anion and a singlet at $\delta = 29.9$ ppm that corresponds to the P–Au phosphorous, which is slightly shifted downfield when compared to the gold complex **165**.

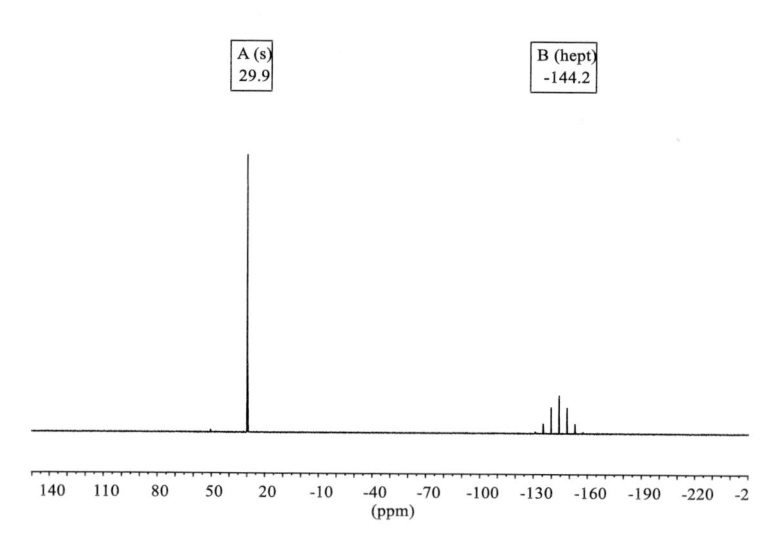

Figure 3.24 ^{31}P NMR of compound **145** showing a singlet (A) and a heptet (B) corresponding to P–Au and the PF_6-anion respectively.

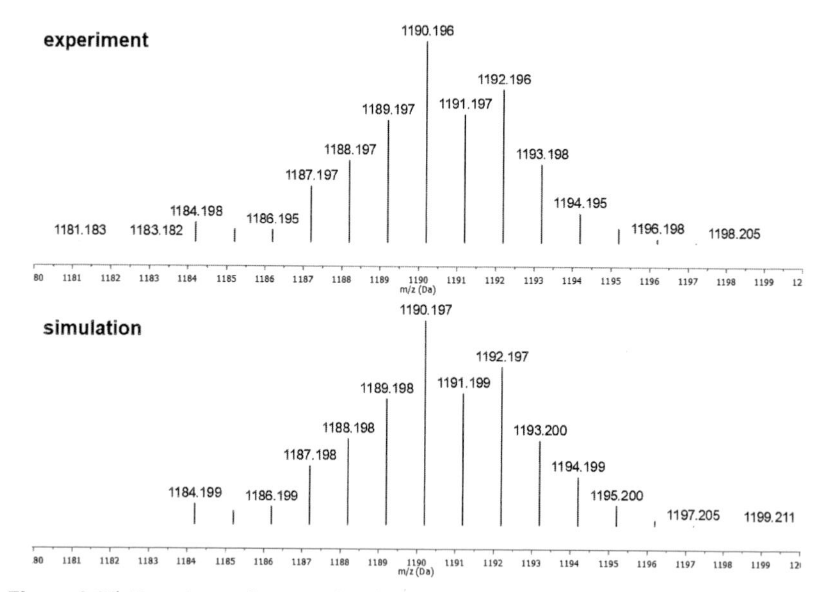

Figure 3.25. Experimental vs. simulated positive mode ESI mass spectrum of **145**.

The obtained ESI MS spectrum of **145** matches very well with the simulated mass spectrum (Figure 3.25). Further CID ESI MS shows fragments of **145** at different energies. The base spectrum shows only fragment **e**, which corresponds to complex **145**. When the energy is raised to 60 eV, two new fragments appear. Fragment **d** corresponds to the formal loss of a bipyridine ligand. The PCP typical homolytic separation of the decks corresponds to the ruthenium complex fragment **a**. When the energy is raised further to 80 eV **a** and **d** become more pronounced. Finally, raising the collision energy to 100 eV leads to fragment **c**, which corresponds to formal loss of two bipyridine ligands and fragment **b**, that additionally shows the formal loss of a chlorine atom.

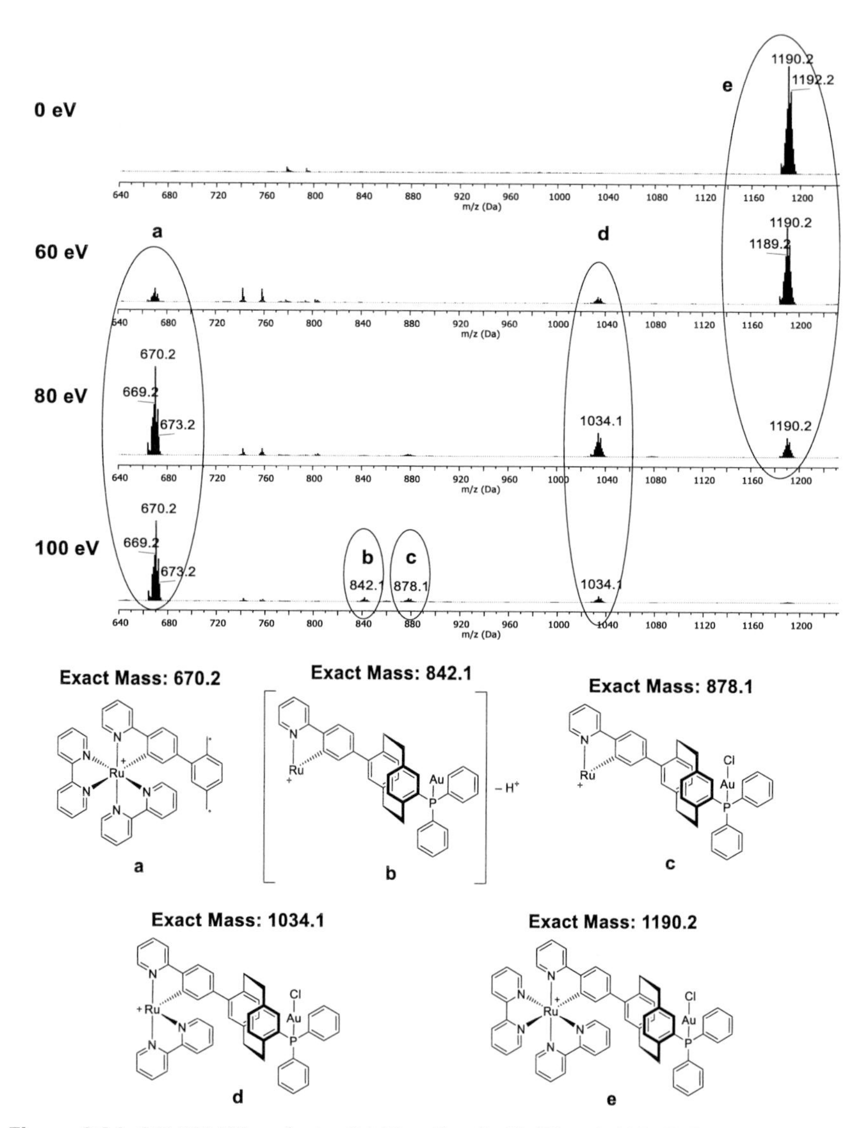

Figure 3.26. CID ESI MS analysis of **145** at E = 0, 60, 80 and 100 eV. Fragments **a-e** correspond to dissociation fragments of **145**.

With this, the synthesis of the target complex **145** was completed in an overall yield of 27% over 5 steps (Scheme 3.78). Synthetic accesses to the other isomers are discussed in the following section.

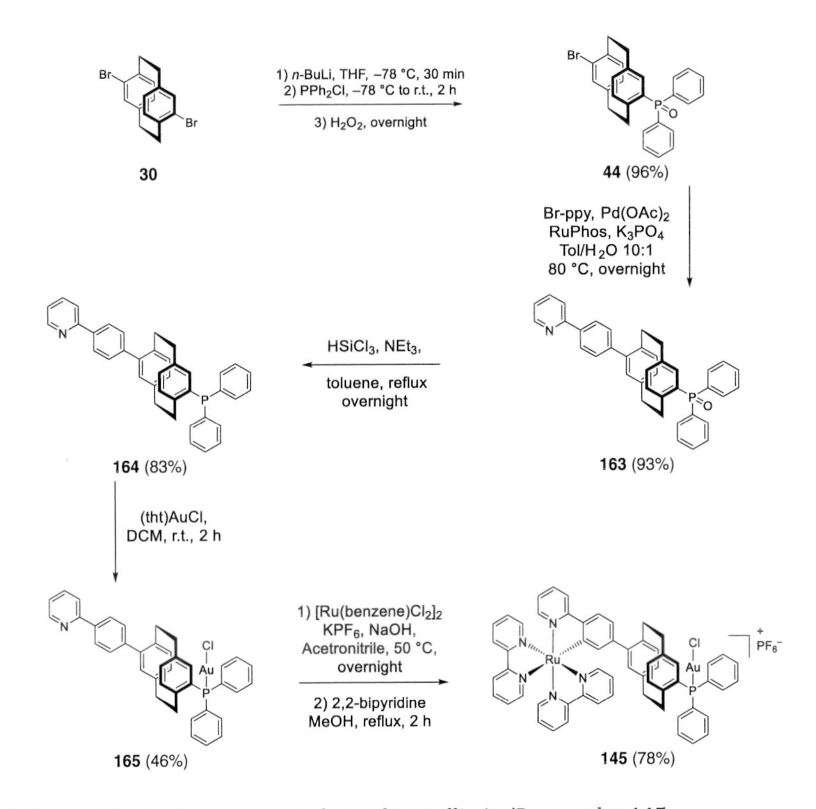

Scheme 3.78. Synthetic access to heterobimetallic Au/Ru complex **145**.

3.6.3 Synthetic Access to other Isomers

After successful synthesis of the pseudo-*para* configured heterobimetallic Au/Ru complex **145**, pseudo-*ortho* and pseudo-*geminal* substitution patterns were aimed for.

3.6.3.1 Pseudo-*geminal* Au-PCP-Ru

The pseudo-geminal substitution pattern is not accessible through the respective dibromide of PCP. Thus, another synthesis strategy had to be devised. Monosubstituted PCPs can be further brominated and interestingly, the first substituent, if hydrogen accepting in nature, directs the bromination to afford the pseudo-*geminal* brominated compound as a major product.[2] Exploiting this behavior, phosphine oxide **175** was subjected to a bromination reaction to afford pseudo-*geminal* brominated phosphine oxide **176** (Scheme 3.79).

Br_2, Fe

DCM, reflux, overnight

175 **176** (84%)

Scheme 3.79. Directed bromination of monophosphine oxide **175** to afford the pseudo-*geminal* phosphine oxide **176**.

However, applying the Suzuki cross-coupling conditions to install the phenylpyridine unit was not successful (Scheme 3.80), which is hypothesized to be due to the sterical encumbrance with pseudo-*geminal* PCP **176**. With the current synthetic tools, no other synthetic access to obtain **177** was possible. The sterics posed here are worse than in the case of pseudo-*ortho* **178**. For this reason, efforts were focused on the pseudo-*ortho* substitution pattern.

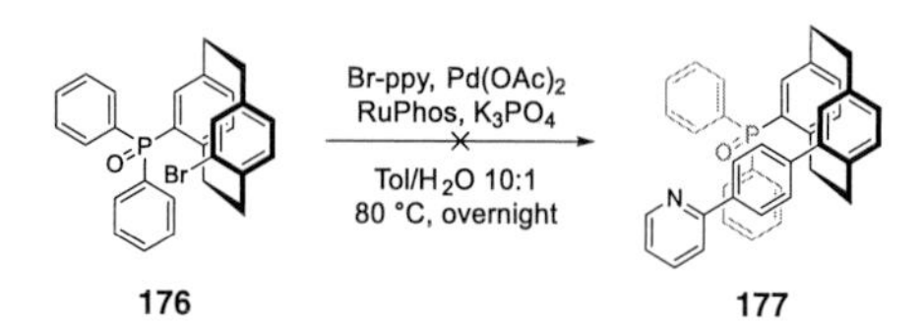

Scheme 3.80. Unsuccessful Suzuki cross-coupling to synthesize pseudo-geminal PCP **177**.

3.6.3.2 Pseudo-*ortho* Au-PCP-Ru

Pseudo-*ortho* PCP derivatives are accessible *via* the pseudo-*ortho* dibromide obtained by thermal isomerization of the pseudo-*para* dibromide of PCP as has been highlighted in section 1.1.

The pseudo-*ortho* dibromide **34** could be converted to the monophosphine oxide **174** analogous to the pseudo-*para* synthesis (Scheme 3.81).

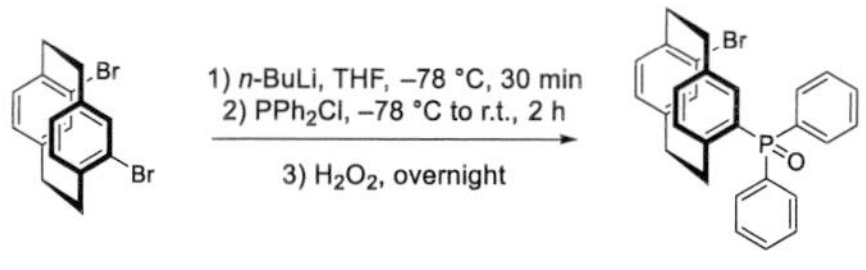

1) *n*-BuLi, THF, –78 °C, 30 min
2) PPh_2Cl, –78 °C to r.t., 2 h
3) H_2O_2, overnight

34 → **178** (72%)

Scheme 3.81. Synthesis of pseudo-*ortho* monophosphine oxide **178**.

However, when trying to attach the phenylpyridine by the established Suzuki cross-coupling reaction with the remaining bromine handle of **178**, the product was only formed in traces according to FAB MS. This might be due to the steric encumbrance of the bromine atom in **178** that does not allow for the formation of the sterically demanding palladium-PCP complex.

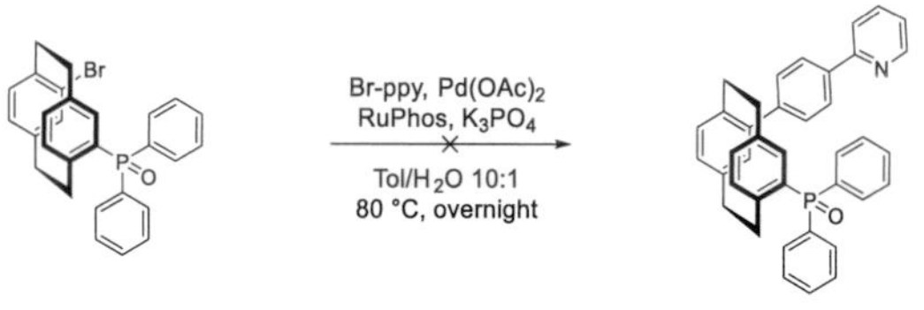

Br-ppy, $Pd(OAc)_2$
RuPhos, K_3PO_4
Tol/H_2O 10:1
80 °C, overnight

178 → **179** (traces)

Scheme 3.82. Unsuccessful synthesis of phenylpyridyl phosphine oxide **179**.

It was reasoned that an inversion regarding the introduction of phosphine oxide and phenylpyridyl substituent could resolve this problem: the lithiation and electrophile trapping to generate the phosphine oxide is sterically far less demanding than the palladium catalysis step. Thus, the trifluoroborate Suzuki cross-coupling chemistry developed in section 3.3 was applied to this challenge. Pseudo-*ortho* halotrifluoroborate **50** could be successfully converted to the phenylpyridyl substituted PCP **180** (Scheme 3.83).

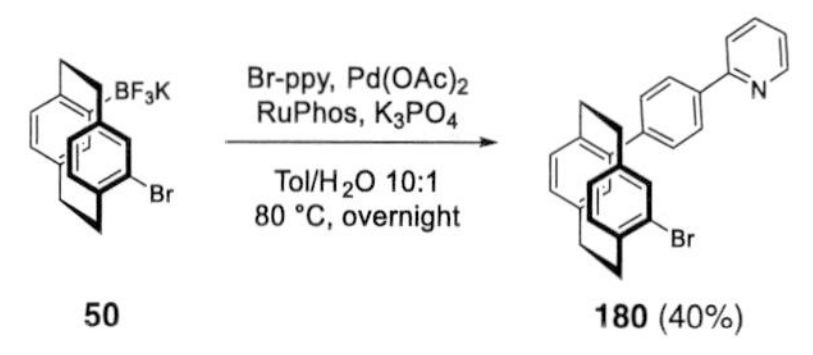

Scheme 3.83. Successful synthesis of phenylpyridyl PCP **180**.

The established three-step one-pot lithiation, phosphination and auration procedure with the remaining bromine handle of **180** however failed to afford the gold(I) complex **181** (Scheme 3.84). Instead, the mixture turned dark brown purple during the auration step, and a precipitate formed. In the working examples above, no color changes or precipitates were observed, thus a reduction to gold(0) of the gold(I) instead of coordination to the phosphine is likely.

N
Br
1) *t*-BuLi, –78 °C, THF, 30 min
2) PPh_2Cl, –78 °C to r.t., THF, 2 h
3) (tht)AuCl, DCM, r.t., 2 h
N
P
Au
Cl
180
181

Scheme 3.84. Unsuccessful synthesis of gold(I) complex **181**.

As this synthetic route also failed, one more cross-coupling approach was tried. The high steric encumbrance of the bromide on **178** was suspected to be responsible for the failing cross-coupling protocol used before. Thus, a ligand known for its high activity in sterically challenging cross-coupling reactions was tried.[180] However, the reaction turned black upon heating and did not deliver the desired product (Scheme 3.85).

Br
P=O
Br-ppy, $Pd(OAc)_2$
AntPhos, K_3PO_4
Tol/H_2O 10:1
80 °C, overnight
N
P=O
O
P
*t*Bu
AntPhos
178
179

Scheme 3.85. Unsuccessful cross-coupling reaction with a new ligand system.

Since a color change towards black indicates the formation of Pd(0) species, the chosen conditions seemed to deteriorate the intermediately formed Pd/BI-DIME complex. Therefore, it was tried to overcome this with a water-free solvent system. Remarkably, a solvent change to dioxane delivered the product in good yield (Scheme 3.86).

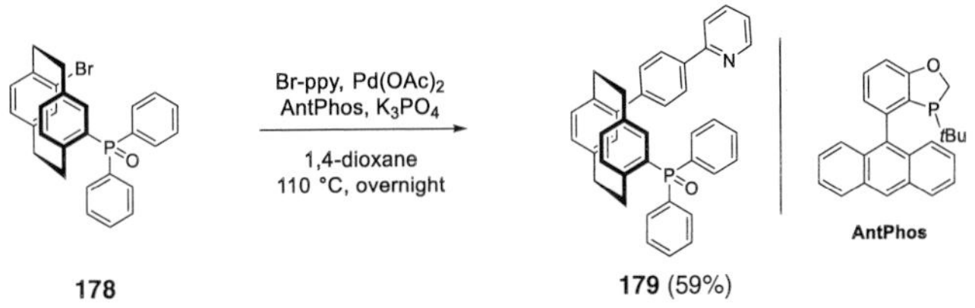

Br-ppy, $Pd(OAc)_2$, AntPhos, K_3PO_4, 1,4-dioxane, 110 °C, overnight

178 → **179** (59%)

AntPhos

Scheme 3.86. Synthesis of phenylpyridyl phosphine oxide **179**.

The subsequent steps were carried out similarly to the synthesis route developed for the pseudo-*para* derivative and afforded the desired heterobimetallic complex **183** after reduction, auration and cycloruthenation in 4% yield over 3 steps (Scheme 3.87). ESI MS and NMR spectroscopy confirmed the identity of pseudo-*ortho* Au/Ru complex **183**.

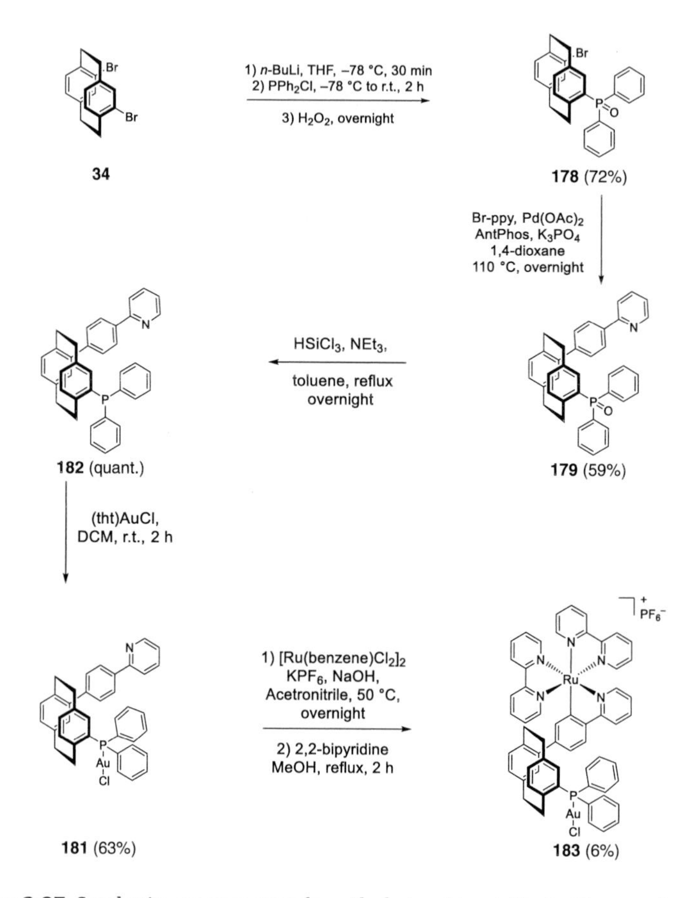

Scheme 3.87. Synthetic access to pseudo-*ortho* heterobimetallic Au/Ru complex **183**.

3.6.4 Application in Photoredox Catalysis

With the heterobimetallic complexes in hand, a suitable test reaction to probe their performance in photoredox catalysis was searched for. A gold(I) and ruthenium(II)-catalyzed, visible-light mediated, arylative Meyer-Schuster rearrangement (MSR) reaction (*vide infra*) was chosen because the gold and ruthenium complexes utilized as catalysts in the published literature[69,70,72] are structurally very close to the heterobimetallic complexes synthesized in section 3.6.2. and 3.6.3.

3.6.4.1 The photoredox-catalyzed Meyer-Schuster rearrangement

The Meyer-Schuster rearrangement is the rearrangement of propargyl alcohols to α,β-unsaturated ketones or aldehydes, catalyzed by acids.[181] The mechanism involves protonation of the propargyl alcohol **184** that can eliminate water, leading to the allene **187**. Attack of water on the allene **187** leads to protonated allenol **188** that after deprotonation tautomerizes to become α,β-unsaturated carbonyl **190**.

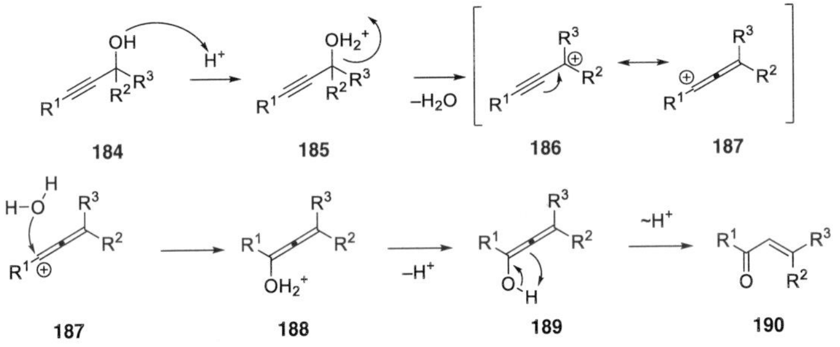

Scheme 3.88. Mechanism of the acid-catalyzed Meyer-Schuster rearrangement.

Besides acids, a range of transition metals such as vanadium,[182,183] titanium,[184] rhenium,[185] molybdenum,[186] ruthenium,[187,188] are efficient catalysts for the MSR. However, gold catalysis promises conversion in mild conditions and both gold(III) and gold(I) catalysts have been reported to catalyze the MSR.[189–192] In the last decade, efforts have been made to expand the utility of the MSR by turning it into part of a domino reaction, e.g. *in situ* functionalization of the alkene **190** *via* Heck cross-coupling.[193] Furthermore, the merging of the MSR with visible-light mediated ruthenium chemistry has seen a surge of interest in recent years.[52,69,70,72] In 2016, the groups of Glorius,[69]

Luna[72] and Shin[70] reported simultaneously on an arylative MSR-catalyzed by gold(I) and ruthenium(II). Through this reaction cascade, α-arylated α,β-unsaturated carbonyls are accessible in a convenient, fast, versatile and mild manner. The mechanism is suggested to involve the following (Scheme 3.89): the coordinatively saturated Au(I) precatalyst **A** undergoes a single electron oxidative addition with an aryl radical to form arylated Au(II) complex **B**. The aryl radical is provided by oxidative quenching of the visible light activated photocatalyst **b** or by single electron oxidation of **B** by the aryl diazonium salt under the extrusion of N_2. **B** can be oxidized to form cationic Au(III) complex **C** by the Ru(III) complex **c** that in turn is reduced back to the catalyst resting state **a**. The coordinatively unsaturated cationic **C** can coordinate to the alkyne of a propargyl alcohol to form π-complex **D**. Deprotonation and the MSR typical 1,3-hydroxyl shift lead to complex **E** that is ready to undergo reductive elimination to fuse the aryl–C bond, thereby affording product **F** and regenerating precatalyst **A**.

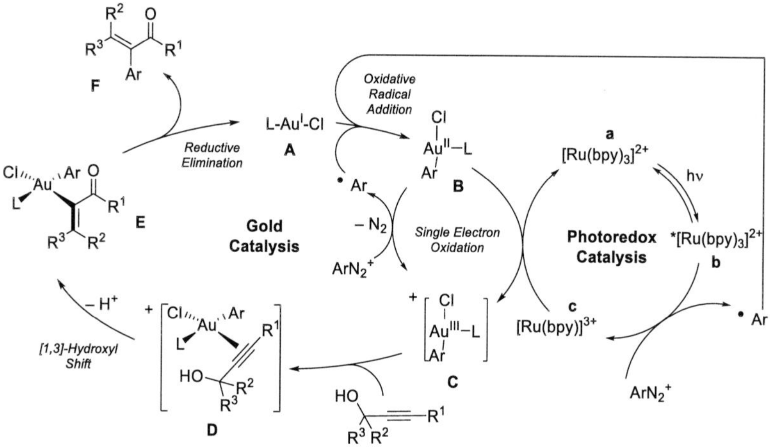

Scheme 3.89. Catalytic Cycle for the gold and ruthenium-catalyzed MSR.

The catalyst **pAuRu** (=**145**) prepared in section 3.6.2 was designed to answer the central research question: Do the metals in a suitably connected heterobimetallic complex communicate with each other and how is this influenced by the distance between the metal centers? Unfortunately, the pseudo-*ortho* isomer **oAuRu** (=**183**) was successfully prepared only towards the end of this work, which is why no photoredox catalysis tests were done with it.

3.6.4.2 Comparability

A few differences between the catalysts used in literature and **pAuRu** are to be kept in mind. Besides (i) replacing a phenyl ring with a sterically demanding PCP scaffold, (ii) the ruthenium(II) has a slightly different coordination environment as there are not three bpy ligands but two bpy ligands and one phenylpyridine ligand in **pAuRu**. Thus, it was necessary to factor in the differences (i) and (ii). For (i) it was planned to prepare the monometallic versions of PPh_3AuCl and $Ru(bpy)_3Cl_2$ with PCP/PCP-phenylpyridine instead of one of the phenyl/bipyridine groups.

To this end, gold(I) complex **191** was synthesized from 4-bromo[2.2]paracyclophane (**51**) in 53% yield by the established one-pot lithiation-phosphination-auration (Scheme 3.90).

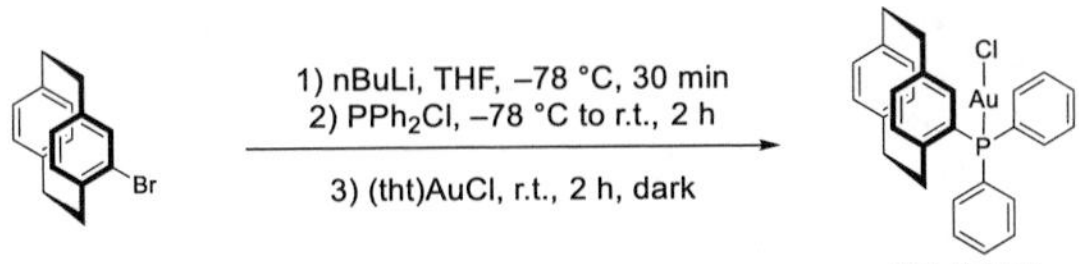

Scheme 3.90. Synthetic access to monometallic PCP gold(I) complex **191**.

To construct the ruthenium coordination environment, monometallic PCP ruthenium complex **192** was envisioned. Since no silver-salt sensitive groups are present, the silver-route can be utilized to obtain PCP ruthenium(II) complex **192** in 46% yield from phenylpyridyl PCP **77** (Scheme 3.91).

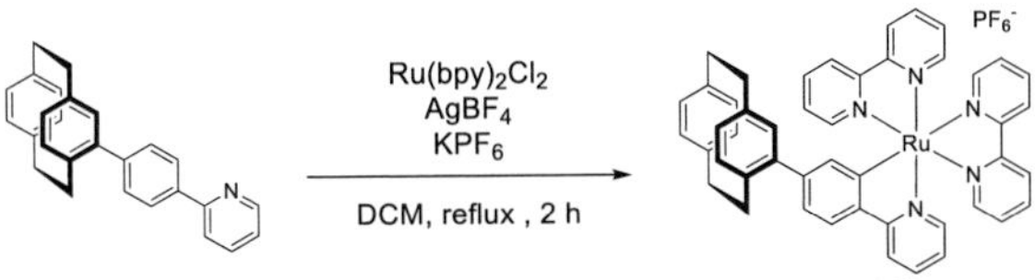

Scheme 3.91. Synthetic access to monometallic PCP ruthenium(II) complex **192**.

3.6.4.3 Absorption Spectroscopy[194]

"Absorption spectra of the complexes PPh_3AuCl (red), $[Ru(bpy)_2(ppy)]PF_6$ (blue), ***PCP–Au*** *(black),* ***PCP–Ru*** *(orange) and* ***pAuRu*** *(green) were obtained in MeOH/MeCN, 3:1 at 298 K*

(Figure 3.27). Absorption maxima in the UV-vis spectra and molar extinction coefficients are summarized in Table 8.

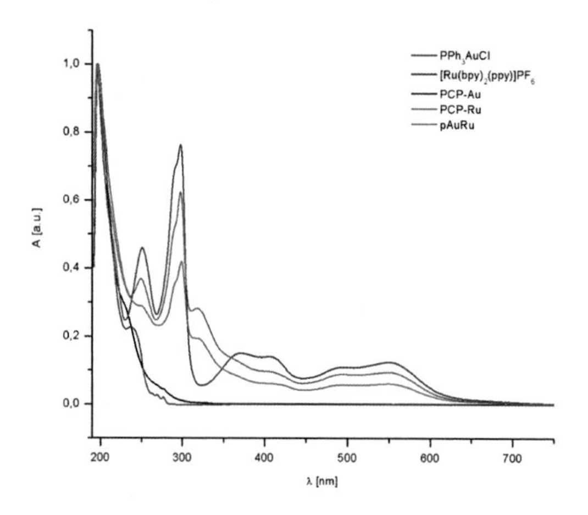

Figure 3.27. Absorption spectra in MeOH/MeCN, 3:1 at 298 K.

Table 8. Absorption spectral properties of the metal complexes in MeOH/MeCN 3:1 at 298 K.

Entry	Compound	λ_{max}^{abs} [nm] (ε [10^4 $M^{-1}cm^{-1}$])
1	[Ru(bpy)$_2$(ppy)]PF$_6$	371 (0.99)
		410 (0.91)
		490 (0.71)
		552 (0.81)
2	PCP–Ru	366 (1.21)
		414 (0.89)
		490 (0.83)
		552 (0.89)
3	pAuRu	366 (0.87)
		414 (0.62)
		490 (0.60)
		552 (0.63)

The absorption bands in the ultraviolet region below 350 nm arise from spin allowed ligand centered (LC) π→π transitions. As the only feature,* ***PCP–Ru*** *and* ***pAuRu*** *exhibit an additional LC band at 320 nm which arises from the PCP backbone. Examination of the lower energy absorption bands reveals a large spectral envelop which arises from the cyclometallated phenylpyridine ligand. This also causes additional features below 300 nm which can be assigned to intraligand π→π* transitions. In the spectra, two bands which are consistent with metal-to-ligand charge transfer (MLCT) are found in the region between 350-450 nm and 450-600 nm. The first band involves population of the excited states involving the cyclometallated ligand, whereas the band at higher wavelengths is primarily localized on the polypyridyl ligands.*

The absorption of ***PCP–Ru*** *and* ***pAuRu*** *is slightly red-shifted compared to [Ru(bpy)*$_2$*(ppy)]PF*$_6$*. The heterobimetallic complex* ***pAuRu*** *also shows lower molar extinction coefficients than the monometallic counterparts."*[194]

3.6.4.4 Emission Spectroscopy

"Excitation of the ***pAuRu*** *complex at wavelengths defined by the absorption maxima at 485 nm and 548 nm produce emission bands that are substantially shifted to longer wavelengths (Figure 3.28). At both excitation wavelengths, strong fluorescence at 795 nm is observed. Upon excitation at 485 nm, the compound shows additional emission at 543 nm and 507 nm."*[194]

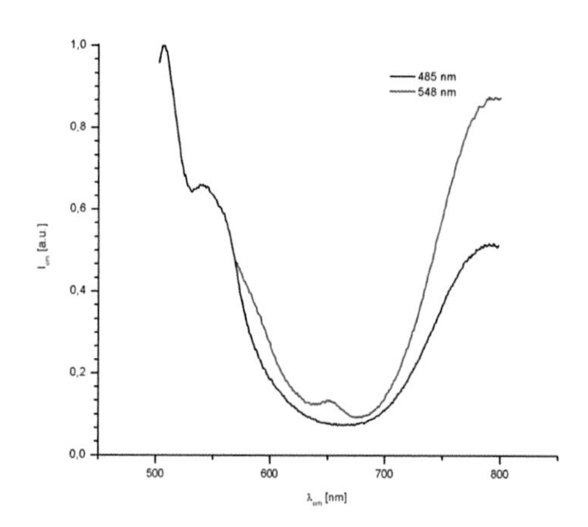

Figure 3.28. Emission spectra at absorption maxima 485 nm (black) and at 548 nm (red) for **pAuRu** in MeOH/MeCN, 3:1 at 298 K.

3.6.4.5 Results

The literature conditions reported by Glorius,[69] Luna[72] and Shin[70] differ considerably in equivalents of catalyst and diazonium salt, solvent composition and yields (Table 9). While Glorius and Shin agree on the necessity of 4.0 equivalents of diazonium salt, Luna reports a surprising trend of an increase in yields going from 4.0 to 2.0 to 1.0 equivalents. At the same time, Glorius reports a significant decrease in yield when using 5 mol% instead of 10 mol% gold catalyst, whereas Shin reports a yield of 80% with only 2.5 mol% gold catalyst (entry 3).

Table 9. Reported literature conditions for the arylative MSR by Glorius,[69] Luna[72] and Shin[70] and results in our hands.

184 → **190**

Ph_3PAuCl **(n mol%)**, $Ru(bpy)_3Cl_2$ **(m mol%)**, ArN_2BF_4 **(z eq.)**, MeOH/ACN **(a:b)**, **1-4 h**, **irradiation**

Entry	group	n	m	z	a	b	Yield [%][a]
1	Glorius	10	2.5	4.0	1	0	72
2	Luna	10	2.5	1.0	3	1	59
3	Shin	2.5	2.5	4.0	1	0	80
4	Bräse	10	2.5	4.0	1	0	71
5	Bräse	10	2.5	4.0	3	1	76
6	Bräse	10	2.5	1.0	3	1	36
7	Bräse	2.5	2.5	4.0	1	0	76

[a] Yields were determined by NMR.

While the solvent did not affect the yield in our case (entry 4 and 5), reducing the amount of diazonium salt (entry 6) lowered the yield significantly. However, a lower catalyst loading of gold at 2.5 mol% (entry 7) did not lead to a decrease in yield. This entry is additionally favorable as it allows for direct comparison with a gold/ruthenium ratio of 1:1, as is given by molecular structure for **pAuRu**. It was thus decided to use the conditions reported by Shin *et al.* for comparison purposes.

The reaction using ethyl carboxylate substituted phenyl diazonium salt **194** served as the definitive model reaction for testing the photoredox catalysis properties of **pAuRu** (Scheme 3.92). The ethyl substituent could be used readily for integration purposes to determine the yield by ^1H NMR.

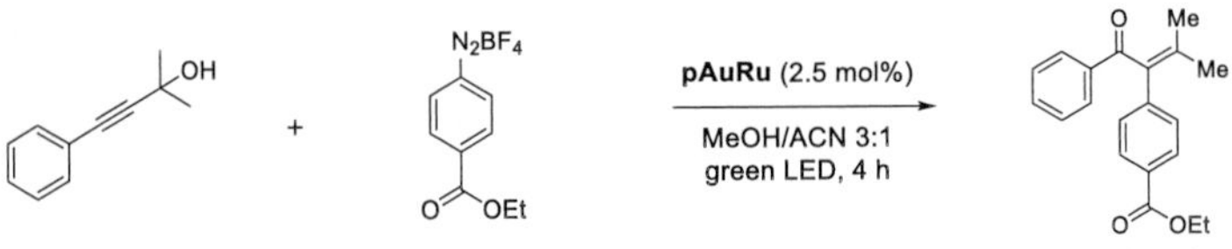

193 (1.00 equiv.) + 194 (4.00 equiv.) → pAuRu (2.5 mol%), MeOH/ACN 3:1, green LED, 4 h → 195 (63%)

Scheme 3.92. Model test reaction for the evaluation of **pAuRu**.

To determine the ability of **pAuRu** to serve as a heterobimetallic photoredox catalyst, a series of reactions were conducted. In a step-wise fashion, the literature conditions were slowly altered to arrive at the **pAuRu** mediated catalysis conditions. With the setup used in our laboratories, the reaction yielded the product **195** in 76% yield (Table 10, entry 1). Because the solubility of the diazonium salt in methanol was very limited, it was first tried to find a better solvent, which was found in acetonitrile/methanol 3:1 (entry 2). Next, the coordination environment for ruthenium was changed accordingly, which did not change the outcome of the catalysis (entry 3). Then, the influence of the PCP scaffold was tested by employing **PCP-Au** and **PCP-Ru** as catalysts. The reaction still delivered the MSR product in 70% yield (entry 4). Finally, **pAuRu** was put to the test and its catalytic activity still formed the product in 63% yield (entry 5). Since electronic and steric effects of the PCP scaffold itself could be excluded by the other experiments, the drop in yield is likely to originate from some degree of communication between the metal centers through the PCP enabled π-π conjugation.

Table 10. Results of the Meyer-Schuster rearrangement with **pAuRu** and the monometallic PCP complexes **PCP-Au** and **PCP-Ru**.

196 (1.00 equiv.) + **197** (4.00 equiv.) → **catalyst** (2.5 mol%), **solvent**, green LED, 4 h → **198**

Entry	catalyst	solvent	Yield [%][a]
1	PPh_3AuCl, $Ru(bpy)_3Cl_2$	MeOH	76[b]
2	PPh_3AuCl, $Ru(bpy)_3Cl_2$	MeCN/MeOH 3:1	73
3	PPh_3AuCl, $Ru(bpy)_2(ppy)Cl_2$	MeCN/MeOH 3:1	73
4	PCP-Au, PCP-Ru	MeCN/MeOH 3:1	70
5	pAuRu	MeCN/MeOH 3:1	63

[a] Isolated yield. [b] NMR yield.

The photoredox catalysis delivered the enone product **195** in a yield of 63%. This strikingly demonstrates the viability of **pAuRu** in photoredox catalysis applications.

3.6.5 Conclusions

Two new heterobimetallic Au/Ru complexes **145** and **183** using PCP as the backbone to hold the metal centers in a fixed spatial relationship were successfully synthesized. The gold(I) is coordinated by a phosphine ligand, while the ruthenium is cyclometallated *via* a phenylpyridine/bis-bipyridyl coordination environment. The preparation of **145** was achieved in a very satisfying yield of 27% over 5 steps. It has been characterized by NMR spectroscopy, HRMS and CID ESI MS. The preparation of **183** demanded for an improved catalyst system and delivered the desired compound in 1.6% yield over 5 steps. Being air- and moisture stable even in solution, **145** and **183** serve as the first of many examples of distance-variable PCP-based heterobimetallic complexes. The access to the remaining pseudo-*geminal* isomer of **145** remains challenging. Additionally, it was shown that complex **145** shows promising activity in a dual Au/Ru photoredox catalysis arylative Meyer-Schuster rearrangement reaction. A slight drop in yield is observed when **145** is employed in the model photoredox catalysis, when compared with the monometallic catalysts. This hints toward a metal-metal interaction, presumably through the π-π conjugation provided by the PCP scaffold.

4 Conclusion and Outlook

The aim of this work was to determine whether PCP is a suitable platform for distance-variable heterobimetallic complexes and show their application in photoredox catalysis.

4.1 Conclusion PCP Trifluoroborates in Suzuki Cross-Coupling

First, a new protocol for the functionalization of PCP by means of Suzuki cross-coupling was developed. The problematic boronic acid of PCP could be replaced by the respective organotrifluoroborate. These compounds are inexpensive, easily prepared on a multi-gram scale, safe and convenient to handle as they are crystalline and bench-stable solids. Moreover, the developed protocol is suitable for the cross-coupling of PCP with a wide range of aryl and vinyl halides in good to excellent yields. Additionally, the class of halotrifluoroborates of PCP was introduced, allowing for the targeted synthesis of heterodisubstituted PCPs in a convenient manner. In conclusion, the regioselective functionalization of PCP has become significantly more convenient and versatile at the same time, while avoiding highly reactive or toxic metal coupling agents (Figure 4.1).

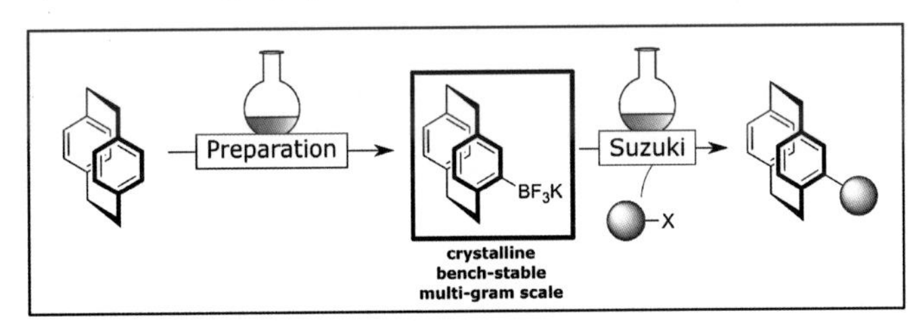

Figure 4.1. Functionalization of PCP with the newly developed PCP trifluoroborate Suzuki cross-coupling protocol.

4.2 Outlook PCP Trifluoroborates in Suzuki Cross-Coupling

PCP trifluoroborates in Suzuki cross-coupling could be expanded by preparation of the enantiopure trifluoroborates by separation of the bromide precursor on a chiral modified HPLC. These enantiopure building blocks could prove very useful in the preparation of chiral PCP containing scaffolds. The same is true for the halotrifluoroborates, where additionally the not yet prepared pseudo-*meta* and pseud-*geminal* isomers are of interest to build more sophisticated PCP scaffolds. Efforts could be put towards the investigation of the low reactivity with halotrifluoroborate PCPs. In addition, the synthesis of unexplored bistrifluoroborates of PCP should be attempted, which are very interesting substrates for the polymerization with aryl halides.

4.3 Conclusion PCP-Porphyrin Conjugates

To combine the advantages of porphyrin systems with the unique features that are offered by PCP, multiple paths towards PCP-porphyrin conjugates and heterobimetallic complexes thereof were pursued.

Initially, a cross-coupling approach was investigated that proved only sufficient to generate homobimetallic complexes. This was due to the necessity to metallate the porphyrin system prior to cross-coupling, thus limiting the flexibility of this approach.

Then, classical condensation reactions were used for the *de novo* synthesis of the porphyrin scaffold. This approach enabled a very flexible synthesis route with regards to the metal atom inserted in the step following the condensation reaction. Additionally, two differing binding sites can be constructed in this way, potentially delivering selectively heterobimetallic complexes instead of the homobimetallic complexes obtained by the cross-coupling approach.

Next, an alkyne-gold zinc-phosphine complex was synthesized and could be detected by HRMS but proved too unstable to isolate. Thus, the more stable cycloruthenated coordination motif was aimed for to increase stability. With this approach, a heterobimetallic Zn/Ru complex could be synthesized, but isolation was not possible as no pure product could be obtained by column chromatography.

4.4 Outlook PCP-Porphyrin Conjugates

The PCP-porphyrins presented here all show strong absorption in the visible-light region. Thus, their application in visible-light mediated photocatalysis could be probed. Also, with the modular synthesis strategy presented, a library of transition metals can be complexed by the porphyrin binding site. By this, all possible spatial relationships provided by the PCP core between the porphyrin complexed metal and the second metal become available, forming a chemical three-dimensional library (Figure 4.2). This library could help in advancing our understanding of metal-metal interactions in heterobimetallic complexes.

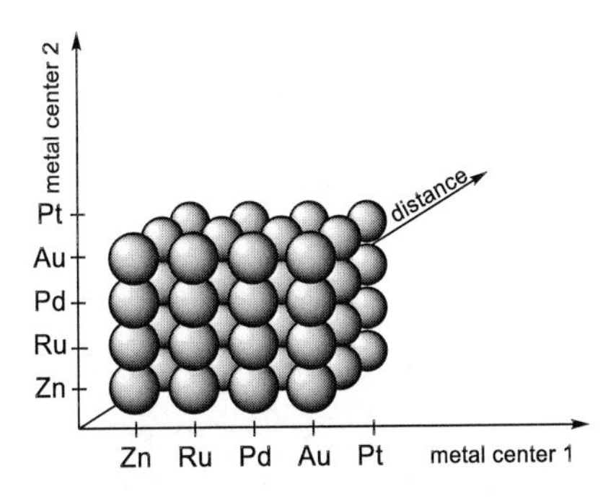

Figure 4.2. Three-dimensional chemical library of heterobimetallic complexes.

4.5 Conclusion Photoredox Catalysis with PCP Au/Ru Complexes

Initially, a first proof-of-principle Au/Ru heterobimetallic complex **130** could be synthesized and characterized by NMR and HRMS (Figure 4.3). However, the elusive compound did not prove to be stable enough for further analysis.

130

Figure 4.3. Proof-of-principle Au/Ru heterobimetallic complex **130**.

To overcome this, heterobimetallic catalyst **pAuRu** was designed and could be synthetically accessed in a convenient 5-step, high-yielding route (Figure 4.4). The dark red solid **pAuRu** shows remarkable air- and moisture stability and the structure was confirmed by NMR, HRMS and CID ESI MS analysis. Additionally, the molecular structure of the direct gold(I) precursor could be elucidated by XRC.

Moreover, the highly interesting pseudo-*ortho* analogue of **pAuRu**, accordingly named **oAuRu**, was accessible with a slight change in synthesis strategy. The proximity of the two metal centers in this molecule is highly interesting for the study of cooperative effects.

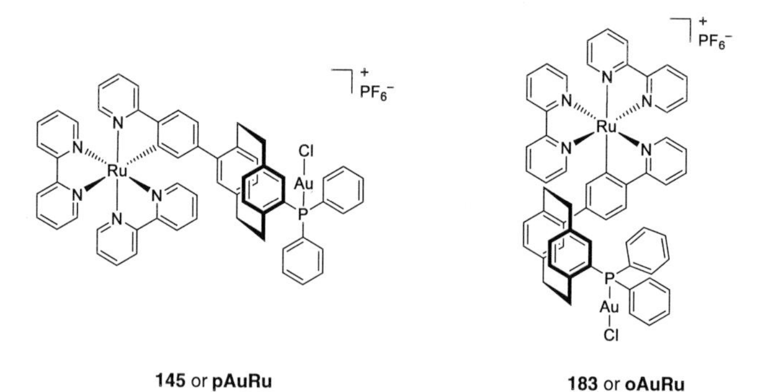

Figure 4.4. Photoredox catalyst **pAuRu** and **oAuRu**.

pAuRu is not only the first-of-its-kind stable and easily accessible heterobimetallic PCP complex, it was also successfully tested for its ability as photoredox catalyst. Here, **pAuRu**

showed promising activity in a dual Au/Ru photoredox catalysis Meyer-Schuster rearrangement, delivering the product in 63% yield with only 2.5 mol% catalyst loading (Figure 4.5).

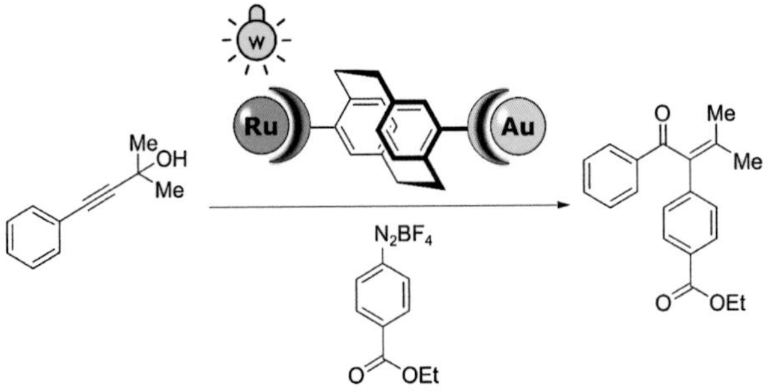

Figure 4.5. A dual Au/Ru photoredox catalysis Meyer-Schuster rearrangement.

4.6 Outlook Photoredox Catalysis with PCP Au/Ru Complexes

The pseudo-*para* derivative of Au/Ru heterobimetallic complex **pAuRu** was successfully prepared and characterized, and additionally displayed great potential in photoredox catalysis. Additionally, pseudo-*ortho* **oAuRu** could be prepared as well.

Future efforts will be geared towards the investigation of **oAuRu** in photoredox catalysis and comparison of these results with the data obtained in this work. Additionally, the metal centers could be variated to host iridium or even less scarce metals as photocatalysts such as copper or iron. The phosphine should also be able to host a palladium atom to open up one of the most researched catalysis fields, i.e. palladium catalysis. Using one of PCPs unique features, the resulting complexes can be prepared in a non-racemic manner to test the amount of chiral induction possible in catalysis applications by the PCP backbone.

After the initial demonstration of **pAuRu**'s ability to serve as a monomolecular heterobimetallic photocatalyst, further investigations towards the reaction order, mechanism and metal-to-metal distance dependency will deliver valuable insights into cooperativity in catalysis.

5 EXPERIMENTAL METHODS

5.1 General Information

Parts of the general information are standardized descriptions and were taken from a previous group member.[195] Literal excerpts thereof are marked by an asterisk *.

5.1.1 Cyclophane Nomenclature

*The IUPAC nomenclature for cyclophanes in general is rather confusing. Therefore Vögtle *et al.* developed a specific cyclophane nomenclature, which is based on a core-substituent ranking.[196] This is exemplified in Figure 5.1 for [2.2]paracyclophane.

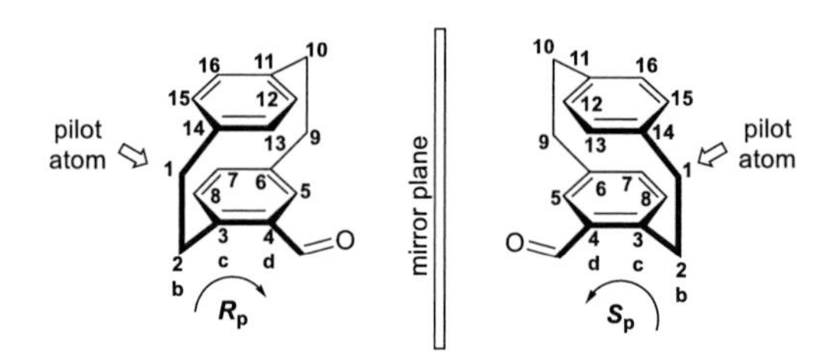

Figure 5.1. Nomenclature illustrated on the two possible enantiomers of 4-formyl-[2.2]paracyclophane.

The core structure is named according to the length of the aliphatic bridges in squared brackets (e.g. [n.m]) and the benzene substitution patterns (*ortho*, *meta* or *para*). [2.2]Paracyclophane belongs to the D_{2h} symmetry, which is broken by the first substituent, resulting in two planar chiral enantiomers. They cannot be drawn in a racemic fashion. By definition, the arene bearing the substituent is set to a chirality plane, and the first atom of the cyclophane structure outside the plane and closest to the chirality center is defined as the "*pilot atom*". If both arenes are substituted, the substituent with higher priority according to the Cahn-Ingold-Prelog (CIP) nomenclature is preferred.[24] The stereo descriptor is determined by the sense of rotation viewed from the pilot atom. To describe the positions of the substituents correctly, an unambiguous numeration is needed. The numbering of the arenes follows the sense of rotation determined by CIP. To indicate the stereochemistry of the planar chirality, a subscripted *p* is added. Unfortunately, the numbering of the second arene is inconsistent in the literature. Therefore, another description based on the benzene substitution patterns is preferred for disubstituted [2.2]paracyclophanes. Substitution on the other ring is commonly named pseudo-(*ortho*, *meta*, *para* or *geminal*). (In this thesis, the rotating direction of numbering of the second ring is set in a way that viewed from the center of the molecule, the sense of rotation consistent, i.e. pseudo-*ortho* is 4,12[2.2]paracyclophane.)

5.1.2 Depiction of Enantiomers/Diastereomers

As none of the reactions in this work were done in an enantio- or diastereoselective manner, all molecules shown are assumed to be mixtures. For convenience and readability, only one arbitrary enantiomer/diastereomer is shown in each case as it is impossible to draw PCP in a racemic manner.

5.1.3 Devices and Analytical Instruments

5.1.3.1 Nuclear Magnetic Resonance Spectroscopy (NMR)

*The NMR spectra of the compounds described herein were recorded on a Bruker Avance 300 NMR instrument at 300 MHz for ^{1}H NMR and 75 MHz for ^{13}C NMR, a Bruker Avance 400 NMR instrument at 400 MHz for ^{1}H NMR, 101 MHz for ^{13}C NMR and 162 MHz for ^{31}P NMR, a Bruker Avance 500 NMR instrument at 500 MHz for ^{1}H NMR, 125 MHz for ^{13}C NMR and 202 MHz for ^{31}P NMR.

The NMR spectra were recorded at room temperature in deuterated solvents acquired from Eurisotop. The chemical shift δ is displayed in parts per million [ppm] and the references used were the ^{1}H and ^{13}C peaks of the solvents themselves: d_1-chloroform ($CDCl_3$): 7.26 ppm for ^{1}H and 77.0 ppm for ^{13}C, d_6-dimethyl sulfoxide (DMSO-d_6): 2.50 ppm for ^{1}H and 39.4 ppm for ^{13}C and d_6-benzene (C_6D_6): 7.16 ppm for ^{1}H and 128.06 ppm for ^{13}C.

For the characterization of centrosymmetric signals, the signal's median point was chosen, for multiplets the signal range. The following abbreviations were used to describe the proton splitting pattern: d = doublet, t = triplet, m = multiplet, dd = doublet of doublet, ddd = doublet of doublet of doublet, dt = doublet of triplet. Absolute values of the coupling constants "*J*" are given in Hertz [Hz] in absolute value and decreasing order. Signals of the ^{13}C spectrum were listed without assigning of the signals.

5.1.3.2 Infrared Spectroscopy (IR)

*The infrared spectra were recorded with a Bruker, IFS 88 instrument. Solids were measured by attenuated total reflection (ATR) method. The positions of the respective transmittance bands are given in wave numbers $\tilde{\upsilon}$ [cm^{-1}] and was measured in the range from 3600 cm^{-1} to 500 cm^{-1}.

5.1.3.3 Mass Spectrometry

Electron ionization (EI) and fast atom bombardment (FAB) experiments were conducted using a Finnigan, MAT 90 (70 eV) instrument, with 3-nitrobenzyl alcohol (3-NBA) as matrix and reference for high resolution. In case of high-resolution measurements, the tolerated error is 0.0005 m/z.

APCI and ESI experiments were recorded on a Q-Exactive (Orbitrap) mass spectrometer (Thermo Fisher Scientific, San Jose, CA, USA) equipped with a HESI II probe to record high resolution. The tolerated error is 5 ppm of the molecular mass.

5.1.3.4 Thin Layer Chromatography

*For the analytical thin layer chromatography, TLC silica plates coated with fluorescence indicator, from Merck (silica gel 60 F254, thickness 0.2 mm) were used. UV-active compounds were detected at 254 nm and 366 nm excitation wavelength with a Heraeus UV-lamp, model Fluotest.

5.1.3.5 Weight Scale

For weightings of solids and liquids, a KERN model ABT 220-4NM was used.

5.1.3.6 Gas Chromatography (GC)

*To determine the degree of conversion, gas chromatograms were recorded on a Bruker 430 GC device equipped with a FactorFour™ VF-5ms (30 m × 0.25 mm × 0.25 mm) capillary column and a flame ionization detector (FID).

5.1.4 Solvents and Reagents

*Solvents of p.a. quality (per analysis) were commercially acquired from Sigma Aldrich, Carl Roth or Acros Fisher Scientific and, unless otherwise stated, used without further purification. Dry solvents were obtained from an mbraun solvent purification system or purchased either from Carl Roth, Acros or Sigma Aldrich (< 50 ppm H_2O over molecular sieves). Solvents were degassed by sparging with inert gas (argon) for 30 minutes. All reagents were commercially acquired from abcr, Acros, Alfa Aesar, Sigma Aldrich, Fluorochem TCI, Chempur, Carbolution or Synchemie, or were available in the group. Unless otherwise stated, all chemicals were used without further purification.

5.2 Synthetic Methods and Characterization Data

General Procedure

*Air- and moisture-sensitive reactions were carried out under argon atmosphere in previously baked out glassware using standard Schlenk techniques. Solid compounds were ground using a mortar and pestle before use, liquid reagents and solvents were injected with plastic syringes and stainless-steel cannula of different sizes, unless otherwise specified.

Reactions at low temperature were cooled using shallow vacuum flasks produced by Isotherm, Karlsruhe, filled with a water/ice mixture for 0 °C, water/ice/sodium chloride for –20 °C or isopropanol/dry ice mixture for –78 °C. For reactions at reflux conditions, the reaction flask was equipped with a reflux condenser and connected to the argon line.

Solvents were evaporated under reduced pressure at 40 °C using a rotary evaporator. Unless otherwise stated, solutions of inorganic salts are saturated aqueous solutions.

Reaction Monitoring

The progress of the reaction in the liquid phase was monitored by TLC. UV active compounds were detected with a UV-lamp at 254 nm and 366 nm excitation wavelength. When required, vanillin solution, potassium permanganate solution or methanolic bromocresol green solution was used as TLC-stain, followed by heating. Additionally, APCI-MS (atmospheric pressure chemical ionization mass spectrometry) was recorded on an Advion expression CMS in positive ion mode with a single quadrupole mass analyzer. The observed molecule ion is interpreted as $[M+H]^+$.

Product purification

Unless otherwise stated, the crude compounds were purified by column chromatography. For the stationary phase of the column, silica gel, produced by Merck (silica gel 60, 0.040 × 0.063 mm, 260 – 400 mesh ASTM), and sea sand by Riedel de-Haën (baked out and washed with hydrochloric acid) were used. Solvents used were commercially acquired in HPLC-grade and individually measured volumetrically before mixing.

5.2.1 Synthetic Methods and Characterization Data for Chapter 3.2

4,16-Dibromo[2.2]paracyclophane (30)

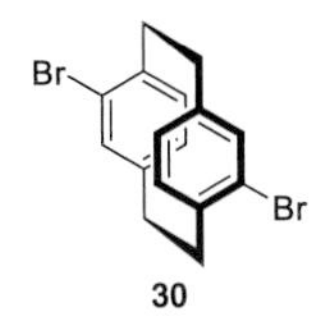

In a 500 mL three-necked flask 241 mg iron powder (4.32 mmol, 4.5 mol-%) was placed. In a dropping funnel a mixture of 10.3 mL bromine (201 mmol, 2.10 equiv.) in 80 mL dichloromethane was placed. Then 15 mL of the mixture was dropped to the iron powder and stirred for 2 hours. After the period the solution was diluted with 100 mL dichloromethane. After addition of 20.0 g [2.2]paracyclophane (96.0 mmol, 1.00 equiv.) and 30 minutes of stirring, the remaining bromine/dichloromethane mixture was added slowly over 5 hours and the resulting brown solution stirred for 3 days. Then, 100 mL of saturated aqueous sodium sulfite was added and stirred until the organic phase went colorless. The precipitate was filtered, washed with water, dissolved in 150 mL boiling toluene, filtered again and placed into a freezer overnight. After anew filtration and drying at 60 °C a white solid was obtained (6.82 g, 19%).

m.p. 252 °C

^{1}H NMR (300 MHz, Chloroform-*d*) δ 7.14 (dd, J = 7.8, 1.9 Hz, 1H), 6.51 (d, J = 1.8 Hz, 1H), 6.44 (d, J = 7.8 Hz, 1H), 3.49 (ddd, J = 12.8, 10.3, 2.0 Hz, 1H), 3.16 (ddd, J = 12.1, 10.2, 4.6 Hz, 1H), 3.02 – 2.78 (m, 2H).

The analytical data is in agreement with the data reported in literature.[88]

4,12-Dibromo[2.2]paracyclophane (34)

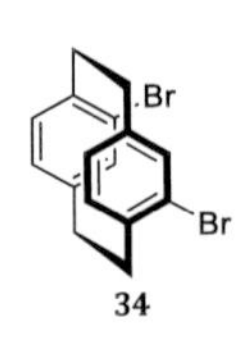

In a 250 mL round bottom flask 4,16-dibromo[2.2]paracyclophane (7.50 g, 20.5 mmol, 1.00 equiv.) was refluxed in triglyme (32.0 mL) for 4 h. After the mixture cooled down, white crystals formed on the bottom. The mixture was filtered and washed with diethyl ether. The residue was subjected to 3 more cycles of refluxing, filtering and washing. The unified filtrates were washed with sat. aq. sodium chloride to remove the remaining triglyme. The solvent was removed under reduced pressure. The crude was subjected to a short silica plug purification (pentane/ethyl acetate 10:1) to yield the title compound as a white solid (5.90 g, 79%).

m.p. 252 °C

^{1}H NMR (400 MHz, Chloroform-*d*) δ 7.20 (d, *J* = 1.6 Hz, 2H), 6.58 – 6.49 (m, 4H), 3.45 (ddd, *J* = 13.3, 9.5, 2.2 Hz, 2H), 3.16 – 2.95 (m, 4H), 2.81 (ddd, *J* = 13.3, 10.1, 6.9 Hz, 2H).

The analytical data is in agreement with the data reported in literature.[102]

4-Bromo-16-diphenylphosphoryl[2.2]paracyclophane (44)[64]

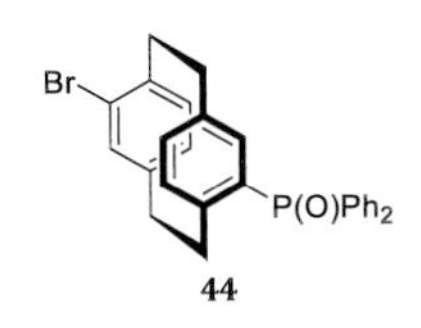

A 500 mL Schlenk flask was charged with 4,16-dibromo[2.2.]-paracyclophane (1.50 g, 4.10 mmol, 1.00 equiv.) and dry THF (100 mL). The reaction mixture was cooled to –78 °C and *n*-BuLi (2.00 mL, 4.92 mmol, 1.20 equiv.) was added dropwise *via* syringe. The solution became pink in color and faded to a pale yellow. The reaction was stirred for 30 min. Then chlorodiphenylphosphine (1.51 mL, 1,81 g, 8.20 mmol, 2.00 equiv.) was added. The mixture was stirred overnight and allowed to warm slowly to room temperature. After the mixture was quenched with a saturated solution of NH_4Cl, the mixture was extracted with ethyl acetate (3×50 mL). The organic layers were washed with brine, dried over $MgSO_4$ and the solvents removed under reduced pressure. The residue was resolved in dichloromethane (100 mL) and H_2O_2 (35%, 7.50 mL, 73.8 mmol, 18.0 equiv.) was added and stirred overnight. Purification *via* column chromatography (silica, CH/EA 3:1) yielded the title compound as a white solid (1.42 g, 96%).

R$_f$ = 0.17 (cyclohexane/ethyl acetate 3:1).

m.p. 211 °C.

^{1}H NMR (400 MHz, $CDCl_3$) δ [ppm] = 7.74 – 7.63 (m, 2H), 7.58 – 7.43 (m, 6H), 7.42 – 7.34 (m, 2H), 7.29 (td, *J* = 7.9, 1.9 Hz, 2H), 6.60 (d, *J* = 1.7 Hz, 1H), 6.55 (dd, *J* = 7.7, 4.2 Hz, 1H), 6.28 – 6.25 (m, 1H), 6.25 – 6.20 (m, 1H), 3.62 – 3.51 (m, 1H), 3.50 – 3.35 (m, 2H), 3.26 – 3.04 (m, 1H), 2.98 – 2.68 (m, 4H).

^{13}C NMR (101 MHz, $CDCl_3$) δ [ppm] = 146.0, 145.9, 142.4, 139.5, 139.4, 138.4, 137.3, 137.2, 137.0, 135.5, 135.4, 134.7, 133., 132.7, 132.3, 132.2, 131.7, 131.6, 131.5, 128.6, 128.5, 128.40, 126.4, 35.4, 35.1, 35.1, 35.0, 33.4.

^{31}P NMR (162 MHz, $CDCl_3$) δ [ppm] = 27.2.

IR (ATR) ṽ = 3043, 1596, 1491, 1478, 1448, 1334, 1311, 1224, 1152, 1120, 1044, 1028, 946, 744, 722, 617, 596, 578, 552, 528, 486, 442, 423, 381.

HRMS ($C_{28}H_{24}BrOP$) calc. 486.0743; found 486.0741.

4-Bromo-16-formyl[2.2]paracyclophane (45)[64]

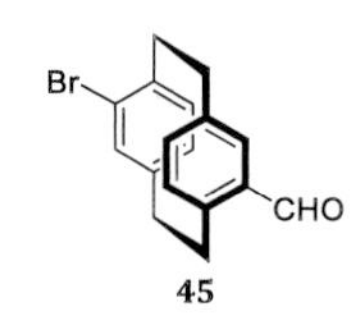

A 500 mL Schlenk flask was charged with 4,16-dibromo[2.2.]paracyclophane (2.50 g, 6.83 mmol, 1.00 equiv.) and dry THF (150 mL). The reaction mixture was cooled to –78 °C and *n*-BuLi (3.28 mL, 8.20 mmol, 1.20 equiv.) was added dropwise *via* syringe. The solution became pink in color and faded to a pale yellow. The reaction was stirred for 30 minutes. Then, dry DMF (4.23 mL, 54.6 mmol, 8.00 equiv.) was added all at once. The mixture was stirred overnight and allowed to warm slowly to room temperature. After the mixture was quenched with a saturated solution of NH_4Cl, the mixture was extracted with ethyl acetate (3×50 mL). The organic layers were washed with brine, dried over Na_2SO_4 and the solvents removed under reduced pressure. Purification *via* column chromatography (silica, CH/EA 50:1) afforded the title compound as a white solid (1.33 g, 62%).

Rf = 0.17 (cyclohexane/ethyl acetate 50:1).

^{1}H NMR (400 MHz, CDCl3) δ [ppm] = 10.0 (s, 1H), 7.4 (dd, *J* = 7.8, 2.0 Hz, 1H), 7.0 (d, *J* = 2.0 Hz, 1H), 6.6 – 6.5 (m, 2H), 6.4 – 6.4 (m, 2H), 4.1 (ddd, *J* = 12.6, 9.9, 2.1 Hz, 1H), 3.5 (ddd, *J* = 13.2, 10.4, 2.7 Hz, 1H), 3.3 (ddd, *J* = 13.2, 10.3, 5.2 Hz, 1H), 3.2 – 2.8 (m, 5H).

^{13}C NMR (101 MHz, $CDCl_3$) δ [ppm] = 192.2, 142.9, 141.6, 140.2, 139.1, 137.3, 137.2, 136.7, 135.3, 134.2, 133.9, 130.8, 127.5, 35.3, 34.3, 33.3, 33.1.

IR (ATR, ṽ) = 2928, 2851, 2730, 1678, 1584,1548, 1474, 1433, 1390, 1283, 1226, 1195, 1142, 1029, 970, 892, 872, 837, 741, 709, 648, 622, 522, 454, 401 cm^{-1}.

HRMS ($C_{17}H_{15}O^{79}Br$) calc. 314.0301; found 314.0300

4-Bromo-12-formyl[2.2]paracyclophane (46)[64]

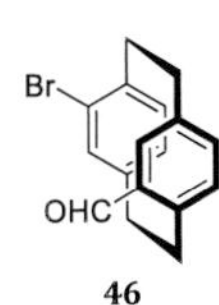

A 500 mL Schlenk flask was charged with 4,12-dibromo[2.2.]paracyclophane (4.00 g, 10.9 mmol, 1.00 equiv.) and dry THF (300 mL). The reaction mixture was cooled to –78 °C and *n*-BuLi (5.25 mL, 13.1 mmol, 1.20 equiv.) was added dropwise *via* syringe. The solution became pink in color and faded to a pale yellow. The reaction was stirred for 30 minutes. Then, dry DMF (6.77 mL, 87.4 mmol, 8.00 equiv.) was added all at once. The mixture was stirred overnight and allowed to warm slowly to room temperature. After the mixture was quenched with a saturated solution of NH_4Cl, the mixture was extracted with ethyl acetate (3×50 mL). The organic layers were washed with brine, dried over Na_2SO_4 and the solvents removed under reduced pressure. Purification *via* column chromatography (silica, CH/EA 50:1) afforded the title compound as a white solid (1.91 g, 56%).

R_f = 0.17 (cyclohexane/ethyl acetate 50:1).

^{1}H NMR (400 MHz, $CDCl_3$) δ [ppm] = 10.0 (s, 1H), 7.4 (dd, *J* = 7.8, 2.0 Hz, 1H), 7.0 (d, *J* = 2.0 Hz, 1H), 6.6 – 6.5 (m, 2H), 6.4 – 6.4 (m, 2H), 4.1 (ddd, *J* = 12.6, 9.9, 2.1 Hz, 1H), 3.5 (ddd, *J* = 13.2, 10.4, 2.7 Hz, 1H), 3.3 (ddd, *J* = 13.2, 10.3, 5.2 Hz, 1H), 3.2 – 2.8 (m, 5H).

^{13}C NMR (101 MHz, $CDCl_3$) δ [ppm] = 192.2, 142.9, 141.6, 140.2, 139.1, 137.3, 137.2, 136.7, 135.3, 134.2, 133.9, 130.8, 127.5, 35.3, 34.3, 33.3, 33.1.

IR (ATR, ṽ) = 2928, 2851, 2730, 1678, 1584,1548, 1474, 1433, 1390, 1283, 1226, 1195, 1142, 1029, 970, 892, 872, 837, 741, 709, 648, 622, 522, 454, 401 cm^{-1}.

HRMS ($C_{17}H_{15}O^{79}Br$) calc. 314.0301; found 314.0300

Ethyl 16-bromo[2.2]paracyclophane-4-carboxylate (47)[64]

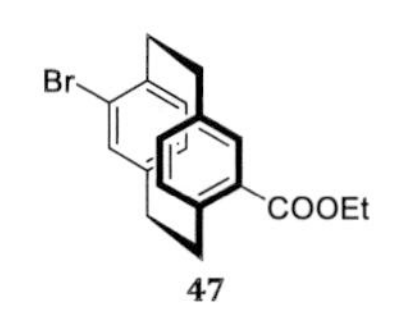

A 500 mL Schlenk flask was charged with 4,16-dibromo[2.2.]paracyclophane (1.50 g, 4.01 mmol, 1.00 equiv.) and dry THF (100 mL). The reaction mixture was cooled to –78 °C and *n*-BuLi (1.97 mL, 4.92 mmol, 1.20 equiv.) was added dropwise *via* syringe. The solution became pink in color and faded to a pale yellow. The reaction was stirred for 1 hour. Then ethyl chloroformate (0.780 mL, 8.20 mmol, 2.00 equiv.) was added all at once. The mixture was stirred overnight and allowed to warm slowly to room temperature. After the mixture was quenched with a saturated solution of NH_4Cl, the mixture was extracted with ethyl acetate (3×50 mL). The organic layers were washed with brine (100 mL), dried over $MgSO_4$ and the solvents removed under reduced pressure. Purification *via* column chromatography (silica, CH/EA 50:1) afforded the title compound as a white solid (880 mg, 60%).

R_f = 0.21 (CH/EA 50:1).

^{1}H NMR (400 MHz, $CDCl_3$) δ [ppm] = 7.32 (dd, *J* = 7.8 Hz *J* = 2.0 Hz, 1H), 7.10 (d, *J* = 2.0 Hz, 1H), 6.66 – 6.48 (m, 4H), 6.42 (d, *J* = 7.6 Hz, 1H), 4.39 (q, *J* = 7.1 Hz, 2H), 4.20 – 3.96 (m, 1H), 3.47 (ddd, *J* = 13.2 Hz *J* = 10.4 Hz *J* = 2.8 Hz, 1H), 3.21 (ddd, *J* = 13.1, *J* = 10.3 Hz *J* =5.0 Hz, 1H), 3.12 – 2.98 (m, 3H), 2.94 – 2.70 (m, 2H), 1.44 (t, *J* = 7.1 Hz, 4H).

The analytical data is in agreement with the data reported in literature.[64]

Potassium 16-bromo[2.2]paracyclophane-4-trifluoroborate (49).

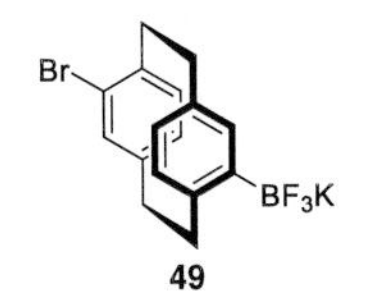

A flame-dried 1 L Schlenk flask was charged with 4,16-dibromo-[2.2.]paracyclophane (3.00 g, 8.20 mmol, 1.00 equiv.) and dry THF (900 mL). The reaction mixture was stirred at room temperature until the starting material was completely dissolved and then was cooled to −78 °C and *n*-BuLi (3.61 mL, 9.01 mmol, 1.10 equiv.) was added dropwise via syringe. The solution became orange in color and faded to a pale yellow. The reaction was stirred for 30 minutes. Then dry trimethyl borate (2.08 mL, 9.01 mmol, 1.10 equiv.) was added all at once. The mixture was stirred for 30 minutes and allowed to warm slowly to room temperature. The next day, sat. aqueous potassium hydrogen fluoride (10.9 mL, 49.2 mmol, 6 equiv.) was added and the mixture stirred for 1 h. The solvents were removed under reduced pressure. The residue was triturated with hot acetone (4 × 100 mL). After removal of the solvent, the residue was washed thoroughly with diethyl ether and dichloromethane and dried under reduced pressure to yield the pure product as a white solid (2.55 g, 79%).

m.p. 280 °C (decomposition).
^{1}H NMR (500 MHz, Acetone-d_6) δ [ppm] = 6.87 (dd, J = 7.6, 1.7 Hz, 1H), 6.80 (dd, J = 7.5, 2.1 Hz, 1H), 6.71 (d, J = 2.1 Hz, 1H), 6.50 (d, J = 1.7 Hz, 1H), 6.33 (d, J = 7.7 Hz, 1H), 6.15 (d, J = 7.6 Hz, 1H), 3.70 (dd, J = 10.7, 8.8 Hz, 1H), 3.32 (ddd, J = 13.1, 10.3, 2.8 Hz, 1H), 3.14 – 3.03 (m, 2H), 2.95 – 2.73 (m, 4H).
^{13}C NMR (101 MHz, Acetone-d_6) δ [ppm] = 143.9, 143.8, 138.9, 137.7, 137.7, 136.9, 136.9, 136.0, 133.7, 132.8, 126.8, 126.5, 36.0, 35.9, 35.3, 34.7.
^{11}B NMR (128 MHz, Acetone-d_6) δ [ppm] −14.9.
^{19}F NMR (376 MHz, Acetone-d_6) δ [ppm] −143.1.
HRMS ($C_{16}H_{14}BBrF_3K$) calc. 391.9961, found 391.9963.

Potassium 12-bromo[2.2]paracyclophane-4-trifluoroborate (50).

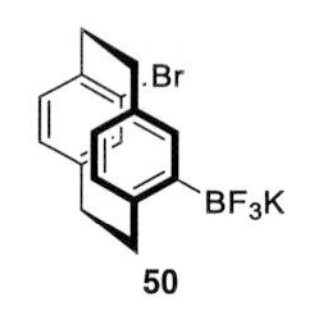

A flame-dried 1 L Schlenk flask was charged with 4,12-dibromo-[2.2.]paracyclophane (3.00 g, 8.20 mmol, 1.00 equiv.) and dry THF (300 mL). The reaction mixture was stirred at room temperature until the starting material was completely dissolved and then was cooled to –78 °C and *n*-BuLi (3.61 mL, 9.01 mmol, 1.10 equiv.) was added dropwise via syringe. The solution became orange in color and faded to a pale yellow. The reaction was stirred for 30 minutes. Then dry trimethyl borate (2.08 mL, 9.01 mmol, 1.10 equiv.) was added all at once. The mixture was stirred for 30 minutes and allowed to warm slowly to room temperature. The next day, sat. aqueous potassium hydrogen fluoride (10.9 mL, 49.2 mmol, 6 equiv.) was added and the mixture stirred for 1 h. The solvents were removed under reduced pressure. The residue was triturated with hot acetone (4 × 100 mL). After removal of the solvent, the residue was washed thoroughly with diethyl ether and dichloromethane and dried under reduced pressure to yield the pure product as a white solid (2.13 g, 66%).

m.p. 325 °C (decomposition).

^{1}H NMR (500 MHz, Chloroform-*d*) δ [ppm] = 7.29 (d, J = 2.1 Hz, 1H), 6.83 (d, J = 1.6 Hz, 1H), 6.54 (dd, J = 7.7, 1.6 Hz, 1H), 6.47 (d, J = 7.7 Hz, 1H), 6.31 (d, J = 7.6 Hz, 1H), 6.23 (dd, J = 7.6, 2.1 Hz, 1H), 3.67 (ddd, J = 12.1, 10.4, 1.8 Hz, 1H), 3.38 – 3.29 (m, 1H), 3.11 – 2.94 (m, 2H), 2.92 – 2.83 (m, 2H), 2.82 – 2.67 (m, 2H).

^{13}C NMR (101 MHz, $CDCl_3$) δ [ppm] = 144.0, 143.5, 139.2, 138.2, 138.2, 138.1, 138.1, 136.8, 135.2, 133.3, 133.3, 133.3, 131.8, 131.5, 127.6, 36.7, 36.6, 35.1, 33.8.

^{11}B NMR (128 MHz, Acetone-*d*$_6$) δ [ppm] -15.0.

^{19}F NMR (376 MHz, Acetone-*d*$_6$) δ [ppm] -144.0.

HRMS ($C_{16}H_{14}BBrF_3K$) calc. 391.9961, found 391.9962.

5.2.2 Synthetic Methods and Characterization Data for Chapter 3.3

4-Bromo[2.2]paracyclophane (51)

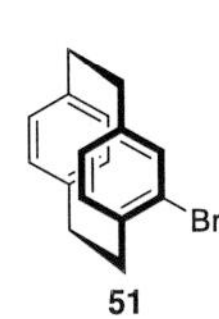

In a 1000 mL three-necked flask iron powder (161 mg, 2.88 mmol, 2 mol%) was placed. In a dropping funnel a mixture of bromine (7.75 mL, 151 mmol, 1.05 equiv.) in dichloromethane (160 mL) was placed. Then, 30 mL of this mixture were dropped to the iron powder and stirred for 2 hours. After the period, the solution was diluted with dichloromethane (500 mL). After addition of [2.2]paracyclophane (30.0 g, 144 mmol, 1.00 equiv.) and 30 minutes of stirring, the remaining bromine/dichloromethane mixture was added slowly over 5 hours and the resulting brown solution stirred overnight. Then, saturated aqueous sodium sulfite (100 mL) was added and stirred until the organic phase went colorless. The mixture was extracted with dichloromethane (3 × 100 mL) and the unified organic phases dried over sodium sulfate. Removal of the solvent under reduced pressure afforded the title compound as a white solid (37.2 g, 90%).

^{1}H NMR (300 MHz, Chloroform-*d*) δ 7.2 (dd, *J* = 7.9, 2.0 Hz, 1H), 6.6 – 6.4 (m, 6H), 3.5 (ddd, *J* = 12.6, 9.9, 2.3 Hz, 1H), 3.3 – 3.0 (m, 5H), 3.0 – 2.8 (m, 2H).

The analytical data is in agreement with the data reported in literature.[102]

Potassium [2.2]paracyclophane-4-trifluoroborate (62)

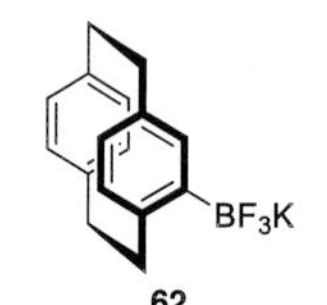

In a round bottom-flask under argon, 4-bromo[2.2]paracyclophane (5.02 g, 17.5 mmol, 1.00 equiv.) was dissolved in 250 mL anhydrous tetrahydrofuran. The solution was cooled to −78 °C and *n*-BuLi (7.70 mL, 2.5 m, 19.3 mmol, 1.10 equiv.) was added dropwise by syringe. After one hour, the yellow solution was quenched with trimethyl borate (6.1 mL, 26.2 mmol, 1.50 equiv.). The now colorless solution was allowed to slowly warm to room temperature. The next day, aqueous potassium hydrogen difluoride (23.3 mL, 4.5 m, 105 mmol, 6.00 equiv.) was added by syringe and the mixture was stirred vigorously for 3 hours. After removal of the solvents under reduced pressure, the white residue was triturated with acetone (2 × room temperature, 2 × boiling, 50 mL each) and the acetone removed under reduced pressure subsequently. The white residue was washed with dichloromethane and diethyl ether (100 mL each) and dried in high vacuum to yield a powdery white crystalline solid (4.79 g, 87%).

m.p. 342 °C (decomposition).

^{1}H NMR (400 MHz, Acetone-d_6) δ [ppm] = 6.76 (dd, *J* = 7.8, 1.5 Hz, 1H), 6.69 (s, 1H), 6.43 (d, *J* = 1.7 Hz, 2H), 6.30 (dd, *J* = 7.8, 1.5 Hz, 1H), 6.17 (d, *J* = 1.7 Hz, 2H), 3.67 (ddd, *J* = 12.5, 10.5, 2.5 Hz, 1H), 3.15 – 3.00 (m, 3H), 2.93 – 2.70 (m, 4H).

^{13}C NMR (101 MHz, Acetone-d_6) δ [ppm] = 144.1, 141.1, 139.6, 137.4, 137.1, 134.7, 134.7, 133.7, 133.6, 132.8, 132.6, 131.2, 36.51, 36.41, 36.33, 36.16.

^{11}B NMR (128 MHz, Acetone-d_6) δ [ppm] −15.2 (d, *J* = 59.2 Hz). – 19F NMR (376 MHz, Acetone-d_6) δ [ppm] −143.23 (m) – IR (ATR) ṽ = 3569, 3378, 2925, 2851, 1894 (vw), 1589, 1552 (vw), 1500 (vw), 1478 (vw), 1436 (vw), 1410, 1330, 1231, 1186 (vw), 1149, 1107, 938, 901, 834, 793, 736, 719, 643, 615 (vw), 590 (vw), 511, 482 (vw).

HRMS (FAB) ($C_{16}H_{15}{}^{11}BF_3K$) calc. 314.0856, found 314.0854.

General ligand-free cross-coupling procedure 3.3A (15, 65-67).

In a vial fitted with a magnetic stirring bar, potassium 4-trifluoroborate-[2.2]paracyclophane (1.25 equiv.), cesium carbonate (4.00 equiv.), palladium acetate (0.05 equiv.) and the respective bromide (1.00 equiv., if solid) were placed. The vial was evacuated and backfilled with argon three times. After addition of the solvent (toluene/water 3:1, 6.00 mL/mmol), the respective bromide (1.00 equiv., if liquid) was added *via* syringe. The mixture was put into a vial heating block and heated to 80 °C for 24 hours. The reaction was cooled to ambient temperature and quenched with sat. aq. ammonium chloride. After separation of the phases, the aqueous phase was extracted with dichloromethane (3 × 15 mL). The organic phases were dried over sodium sulfate and the solvent was removed under reduced pressure. The crude product was purified *via* column chromatography (silica, cyclohexane/ethyl acetate).

General RuPhos mediated cross-coupling procedure 3.3B (15, 68-82).

In a vial fitted with a magnetic stirring bar, potassium 4-trifluoroborate[2.2]paracyclophane (1.50 equiv.), potassium phosphate (4.00 equiv.), palladium acetate (0.05 equiv.), RuPhos (0.15 equiv.) and the respective bromide (1.00 equiv., if solid) were placed. The vial was capped, evacuated and backfilled with argon three times. After addition of the solvent (toluene/water 10:1, 0.1 M), the respective bromide (1.00 equiv., if liquid) was added *via* syringe. The vial was put into a vial heating block and heated to 80 °C for 24 hours. The reaction was cooled to ambient temperature and quenched with sat. aq. ammonium chloride. After separation of the phases, the aqueous phase was extracted with dichloromethane (3 × 15 mL). The organic phases were dried over sodium sulfate and the solvent was removed under reduced pressure. The crude product was purified *via* column chromatography (silica, pentane/ethyl acetate).

4-(2'-Pyridyl)[2.2]paracyclophane (15)

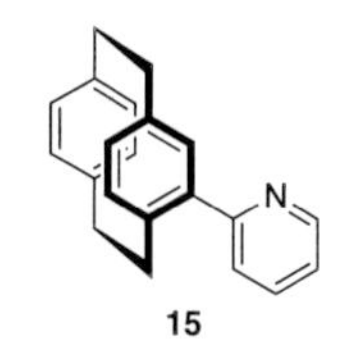

Method A:

According to general procedure **3.3A**, the title compound was obtained by column chromatography as an off-white solid (77 mg, 43%).

Method B:

According to general procedure **3.3B**, the title compound was obtained by column chromatography as an off-white solid (76 mg, 42%).

Method C:

According to general procedure **3.3B** with using toluene/water 1:1 and sodium carbonate as base instead of potassium phosphate (71%).

$\mathbf{R_f}$ = 0.25 (cyclohexane/ethyl acetate 20:1).

^{1}H NMR (400 MHz, $CDCl_3$) δ [ppm] = 8.71 – 8.64 (m, 1H), 7.68 (td, J = 7.7, 1.9 Hz, 1H), 7.43 (dd, J = 8.0, 1.1 Hz, 1H), 7.23 – 7.09 (m, 1H), 6.75 (d, J = 1.9 Hz, 1H), 6.57 – 6.37 (m, 6H), 3.67 – 3.52 (m, 1H), 3.19 – 3.01 (m, 2H), 3.04 – 2.79 (m, 4H), 2.64 – 2.51 (m, 1H).

^{13}C NMR (101 MHz, $CDCl_3$) δ [ppm] = 159.1 ($C_{quat.}$), 149.7 (+, $C_{Ar}H$), 140.6 ($C_{quat.}$), 139.8 ($C_{quat.}$), 139.6 ($C_{quat.}$), 139.4 ($C_{quat.}$), 138.2 ($C_{quat.}$), 136.3 (+, $C_{Ar}H$), 136.2 (+, $C_{Ar}H$), 133.2 (+, $C_{Ar}H$), 132.9 (+, $C_{Ar}H$), 132.7 (+, $C_{Ar}H$), 132.6 (+, $C_{Ar}H$), 132.4 (+, $C_{Ar}H$), 130.7 (+, $C_{Ar}H$), 124.2 (+, $C_{Ar}H$), 121.4 (+, $C_{Ar}H$), 35.57 (–, CH_2), 35.34 (–, CH_2), 35.21 (–, CH_2), 34.59 (–, CH_2).

IR (ATR) $\tilde{\nu}$ = 2930 (vw), 2850 (vw), 1582 (vw), 1498 (vw), 1465 (vw), 1422 (vw), 1239 (vw), 1145 (vw), 1087 (vw), 1037 (vw), 991 (vw), 943 (vw), 893. (vw), 849 (vw), 784 (vw), 747 (vw), 729 (vw), 717 (vw), 653 (vw), 639 (vw), 618 (vw), 595 (vw), 559 (vw), 515 (vw), 481 (vw), 408 (vw).

HRMS ($C_{21}H_{19}N$) calc. 285.1517, found 285.1519.

The analytical data is in agreement with the data reported in literature.[102]

4-(3'-Pyridyl)[2.2]paracyclophane (65)

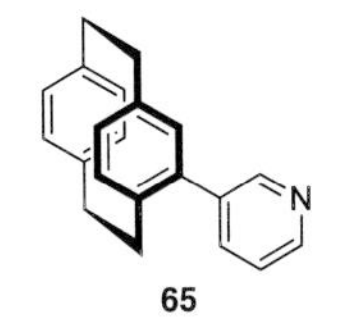

According to general procedure **3.3A**, the title compound was obtained by column chromatography as an off-white solid (126 mg, 70%).

R$_f$ = 0.27 (cyclohexane/ethyl acetate 3:1).
^{1}H NMR (400 MHz, $CDCl_3$) δ [ppm] = 8.77 (d, *J* = 2.2 Hz, 1H), 8.62 (dd, *J* = 4.9, 1.7 Hz, 1H), 7.79 (dt, *J* = 7.7, 2.0 Hz, 1H), 7.40 (dd, *J* = 7.8, 4.8 Hz, 1H), 6.64 (t, *J* = 8.3 Hz, 2H), 6.57 (dd, *J* = 11.1, 1.6 Hz, 5H), 3.36 (ddd, *J* = 13.0, 10.0, 3.1 Hz, 1H), 3.22 – 3.11 (m, 3H), 3.09 – 3.02 (m, 1H), 3.01 – 2.94 (m, 1H), 2.90 (ddd, *J* = 13.5, 10.3, 3.2 Hz, 1H), 2.64 (ddd, *J* = 13.1, 10.0, 4.8 Hz, 1H).

^{13}C NMR (101 MHz, $CDCl_3$) δ [ppm] = 150.5 (+, C_{Ar}H), 147.9 ($C_{quat.}$), 140.2 ($C_{quat.}$), 139.6 ($C_{quat.}$), 139.5 ($C_{quat.}$), 138.2 ($C_{quat.}$), 137.2 ($C_{quat.}$), 137 (+, C_{Ar}H), 136.6 (+, C_{Ar}H), 136.1 (+, C_{Ar}H), 136.1 (+, C_{Ar}H), 133.3 (+, C_{Ar}H), 133.1 (+, C_{Ar}H), 132.8 (+, C_{Ar}H), 132.1 (+, C_{Ar}H), 132 (+, C_{Ar}H), 129.6 (+, C_{Ar}H), 123.5 (–, CH_2), 35.5 (–, CH_2), 35.2 (–, CH_2), 34.9 (–, CH_2).
IR (ATR) $\tilde{\nu}$ = 3027 (vw), 2924 (w), 2849 (w), 1591 (vw), 1567 (vw), 1498 (vw), 1474 (w), 1433 (vw), 1412 (w), 1339 (vw), 1185 (vw), 1109 (vw), 1053 (vw), 1018 (w), 942 (vw), 900 (w), 848 (w), 807 (m), 732 (w), 715 (m), 654 (m), 639 (w), 620 (vw), 594 (w), 514 (m), 478 (m), 403 (vw).
HRMS ($C_{21}H_{19}N$) calc. 285.1517, found 285.1517.

The analytical data is in agreement with the data reported in literature.[102]

4-(4'-Pyridyl)[2.2]paracyclophane (66)

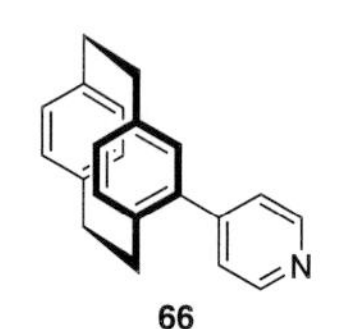

According to general procedure **3.3A**, the title compound was obtained by column chromatography as an off-white solid (122 mg, 68%).

R$_f$ = 0.22 (cyclohexane/ethyl acetate 3:1).

^{1}H NMR (400 MHz, Chloroform-d) δ 8.73 – 8.69 (m, 2H), 7.44 – 7.38 (m, 2H), 6.68 – 6.49 (m, 7H), 3.39 (ddd, J = 12.5, 10.0, 2.9 Hz, 1H), 3.22 – 2.84 (m, 6H), 2.66 (ddd, J = 13.1, 10.0, 4.5 Hz, 1H).

^{13}C NMR (101 MHz, $CDCl_3$) δ [ppm] = 149.7 (+, $C_{Ar}H$), 148.9 ($C_{quat.}$), 140.2 ($C_{quat.}$), 139.6 ($C_{quat.}$), 139.5 ($C_{quat.}$), 139.1 ($C_{quat.}$), 137.3 ($C_{quat.}$), 136.3 (+, $C_{Ar}H$), 133.6 (+, $C_{Ar}H$), 133.3 (+, $C_{Ar}H$), 132.8 (+, $C_{Ar}H$), 132.0 (+, $C_{Ar}H$), 132.0 (+, $C_{Ar}H$), 129.8 (+, $C_{Ar}H$), 124.6 (+, $C_{Ar}H$), 35.55 (+, $C_{Ar}H$), 35.28 (+, $C_{Ar}H$), 35 (–, CH_2), 33.98 (–, CH_2), 35.21 (–, CH_2), 34.59 (–, CH_2).

IR (ATR) $\tilde{\nu}$ = 2923 (m), 2849 (w), 1591 (w), 1499 (w), 1474 (w), 1433 (w), 1411 (m), 1339 (w), 1185 (w), 1106 (vw), 1053 (vw), 1018 (m), 942 (w), 900 (m), 848 (m), 808 (m), 732 (m), 715 (m), 653 (m), 639 (m), 593 (m), 514 (m), 479 (m), 404 (w), 1899 (vw).

HRMS ($C_{21}H_{19}N$) calc. 285.1517, found 285.1519.

The analytical data is in agreement with the data reported in literature.[102]

4-(5'-Pyrimidyl)[2.2]paracyclophane (67)

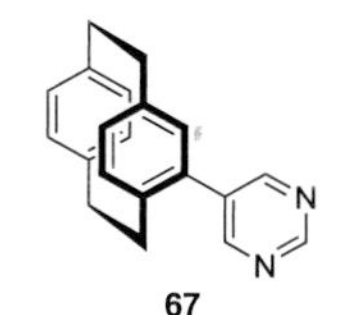

According to general procedure **3.3A**, the title compound was obtained by column chromatography as an off-white solid (1.18 g, 82%).

R$_f$ = 0.29 (cyclohexane/ethyl acetate 3:1).

^{1}**H NMR** (300 MHz, chloroform-*d*) δ 9.23 (d, J = 1.6 Hz, 1H), 8.87 (d, J = 1.5 Hz, 2H), 6.77 – 6.36 (m, 7H), 3.37 – 3.24 (m, 1H), 3.20 – 2.87 (m, 6H), 2.64 (ddd, J = 15.1, 9.9, 4.0 Hz, 1H).

HRMS ($C_{20}H_{18}N_2$) calc. 286.1470, found 286.1471.

The analytical data is in agreement with the data reported in literature.[102]

Methyl 4-([2.2]paracyclophanyl)benzoate (68)

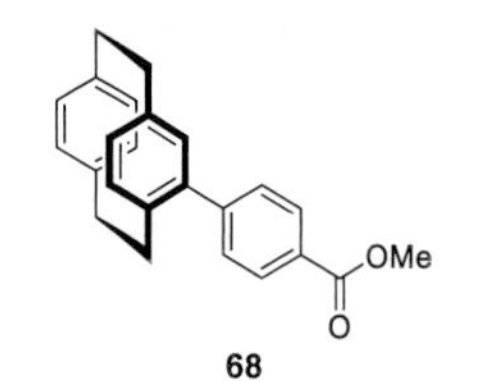

According to general procedure **3.3B**, the title compound was obtained by column chromatography as a white solid (202 mg, 72%).

R$_f$ = 0.08 (pentane/ethyl acetate 50:1).

^{1}H NMR (500 MHz, Chloroform-*d*) δ [ppm] 8.18 (d, J = 8.3 Hz, 2H), 7.58 (d, J = 8.4 Hz, 2H), 6.72 – 6.48 (m, 7H), 3.99 (s, 3H), 3.42 (ddd, J = 12.5, 10.0, 3.0 Hz, 1H), 3.25 – 2.84 (m, 6H), 2.64 (ddd, J = 13.0, 10.0, 4.5 Hz, 1H).

^{13}C NMR (101 MHz, $CDCl_3$) δ [ppm] = 166.1, 144.7, 139.7, 138.9, 138.6, 138.4, 136.2, 135.0, 132.1, 131.8, 131.6, 131.1, 130.9, 128.8 (2C), 128.7, 128.6 (2C), 127.4, 51.1, 34.4, 34.2, 33.8, 33.1.

IR (ATR) $\tilde{\nu}$ = 2923 (w), 1716 (m), 1606 (w), 1503 (vw), 1434 (w), 1313 (w), 1278 (m), 1192 (w), 1176 (w), 1105 (m), 1018 (w), 913 (w), 859 (w), 775 (m), 738 (w), 708 (m), 647 (w), 631 (w), 591 (w), 553 (w), 521 (w), 499 (w), 481 (w), 384 (vw). **HRMS** ($C_{24}H_{22}O_2$) calc. 342.1620, found 342.1620.

4-(*p*-Acetylphenyl)[2.2]paracyclophane (69)

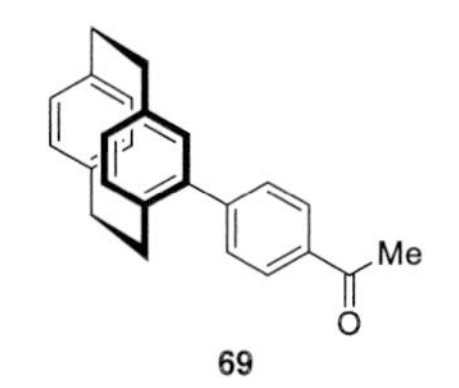

According to general procedure **3.3B**, the title compound was obtained by column chromatography as a white solid (163 mg, 81%).

R$_f$ = 0.08 (pentane/ethyl acetate 50:1).

^{1}H NMR (500 MHz, Chloroform-*d*) δ [ppm] = 8.07 (d, J = 8.4 Hz, 2H), 7.58 (d, J = 8.3 Hz, 2H), 6.69 – 6.49 (m, 7H), 3.40 (ddd, J = 12.8, 10.0, 3.1 Hz, 1H), 3.23 – 2.84 (m, 6H), 2.68 (s, 3H), 2.66 – 2.59 (m, 1H).

^{13}C NMR (101 MHz, $CDCl_3$) δ [ppm] = 198.1, 146.1, 140.8, 140.1, 139.7, 139.6, 137.4, 136.2, 135.5, 133.3, 133.1, 132.8, 132.2, 132.1, 130.0, 129.9, 128.8, 35.6, 35.3, 35.0, 34.3, 26.8.

IR (ATR) $\tilde{\nu}$ = 2926 (w), 2851 (vw), 1673 (m), 1600 (w), 1502 (vw), 1414 (w), 1353 (w), 1306 (vw), 1267 (w), 1177 (w), 1092 (vw), 1015 (vw), 955 (w), 912 (w), 847 (w), 831 (w), 795 (w), 736 (vw), 716 (w), 674 (vw), 650 (w), 632 (w), 595 (w), 585 (w), 520 (w), 495 (w), 409 (vw).

HRMS ($C_{24}H_{22}O_1$) calc. 326.1671, found 326.1669.

The analytical and spectroscopical data match those reported in literature.[197]

4-(4'-Nitrophenyl)[2.2]paracyclophane (70)

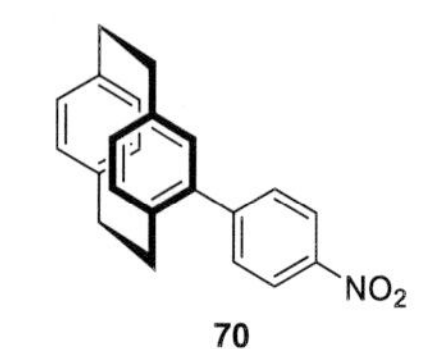

According to general procedure **3.3B**, the title compound was obtained by column chromatography as a white solid (208 mg, 87%).

R_f = 0.30 (pentane/ethyl acetate 50:1).

^{1}H NMR (400 MHz, Chloroform-*d*) δ [ppm] = 8.34 (d, J = 8.8 Hz, 2H), 7.63 (d, J = 8.8 Hz, 2H), 6.75 – 6.37 (m, 7H), 3.35 (ddd, J = 13.1, 10.0, 3.2 Hz, 1H), 3.27 – 3.13 (m, 3H), 3.12 – 2.85 (m, 3H), 2.64 (ddd, J = 13.1, 10.0, 4.7 Hz, 1H).

^{13}C NMR (101 MHz, $CDCl_3$) δ [ppm] 147.9, 146.8, 140.4, 139.7, 139.7, 139.6, 137.4, 136.4, 133.7, 133.5, 132.9, 132.2, 132.0, 130.5 (2C), 129.7, 124.0 (2C), 35.6, 35.3, 35.1, 34.2.

IR (ATR) $\tilde{\nu}$ = 2927, 2890, 2873, 2850, 1591, 1513, 1475, 1436, 1414, 1344, 1315, 1288, 1238, 1179, 1164, 1105, 1091, 1045, 1013, 907, 881, 851, 841, 822, 806, 796, 756, 735, 732, 718, 701, 686, 659, 647, 625, 591, 554, 533, 516, 476, 450, 433, 426, 416, 380.

HRMS ($C_{22}H_{20}O_2N_1$) calc.330.1494, found 330.1495.

4-(3,4,5-Trifluorophenyl)[2.2]paracyclophane (71)

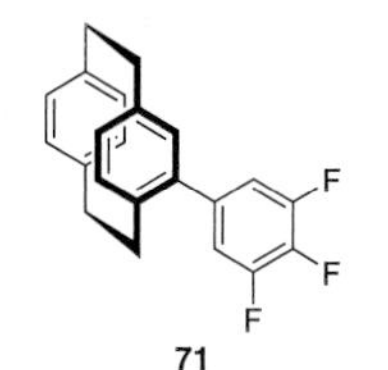

According to general procedure **3.3B**, the title compound was obtained by column chromatography as a white solid (76 mg, 74%).

R$_f$ = 0.34 (pentane/ethyl acetate 200:1).

^{1}H NMR (400 MHz, Chloroform-*d*) δ [ppm] = 7.08 (dd, J = 8.7, 6.6 Hz, 2H), 6.65 (dd, J = 8.2, 1.7 Hz, 1H), 6.63 – 6.51 (m, 5H), 6.46 (d, J = 1.8 Hz, 1H), 3.41 – 3.33 (m, 1H), 3.23 – 3.12 (m, 3H), 3.10 – 2.86 (m, 3H), 2.76 – 2.65 (m, 1H).

^{13}C NMR (101 MHz, $CDCl_3$) δ [ppm] = 152.6 (dd, J = 9.9, 4.4 Hz), 150.1 (dd, J = 9.9, 4.3 Hz), 140.4, 139.6, 137.0, 136.3, 133.4, 133.3, 132.9, 132.0, 131.9, 113.69 – 113.27 (m), 129.6, 35.6, 35.3, 35.0, 34.1.

IR (ATR) $\tilde{\nu}$ = 1616, 1526, 1500, 1486, 1434, 1401, 1353, 1239, 1041, 902, 882, 858, 829, 815, 792, 749, 737, 720, 704, 694, 667, 650, 636, 620, 575, 513, 503, 480.

HRMS ($C_{22}H_{17}F_3$) calc. 338.1282, found 338.1281.

4-(4-Cyanophenyl)[2.2]paracyclophane (72)

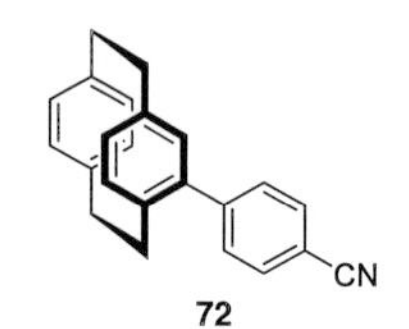

According to general procedure **3.3B**, the title compound was obtained by column chromatography as a white solid (185 mg, 92%).

R$_f$ = 0.29 (pentane/ethyl acetate 50:1).

^{1}H NMR (500 MHz, Chloroform-*d*) δ [ppm] = 7.80 – 7.71 (m, 2H), 7.62 – 7.55 (m, 2H), 6.69 – 6.46 (m, 7H), 3.33 (ddd, *J* = 12.7, 10.0, 3.0 Hz, 1H), 3.24 – 3.10 (m, 3H), 3.11 – 2.84 (m, 3H), 2.64 (ddd, *J* = 13.1, 10.0, 4.5 Hz, 1H).

^{13}C NMR (101 MHz, $CDCl_3$) δ [ppm] = 145.9, 140.3, 140.1, 139.7, 139.6, 137.3, 136.3, 133.5, 133.4, 132.9, 132.5 (2C), 132.2, 132.0, 130.4 (2C), 129.7, 119.2, 110.6, 35.6, 35.3, 35.1, 34.1.

IR (ATR) $\tilde{\nu}$= 2924, 2873, 2850, 2227, 1604, 1591, 1500, 1476, 1435, 1414, 1392, 1367, 1354, 1238, 1108, 1092, 1017, 941, 907, 881, 850, 837, 822, 795, 775, 765, 755, 737, 718, 691, 676, 671, 654, 639, 595, 567, 530, 509, 494, 473, 455, 449, 414, 402, 388, 380.

HRMS ($C_{23}H_{20}N_1+H^+$) calc. 310.1596, found 310.1594.

The analytical and spectroscopical data match those reported in literature.[106]

4-(4'-Tolyl)[2.2]paracyclophane (73)

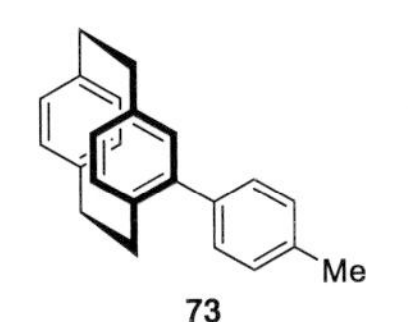

According to general procedure **3.3B**, the title compound was obtained by column chromatography as a white solid (228 mg, 91%).

R_f = 0.21 (pentane/ethyl acetate 200:1).

^{1}H NMR (500 MHz, Chloroform-*d*) δ [ppm] = 7.46 (d, J = 7.9 Hz, 2H), 7.36 (d, J = 7.8 Hz, 2H), 6.78 – 6.67 (m, 1H), 6.68 – 6.61 (m, 4H), 6.57 (dd, J = 7.8, 1.9 Hz, 2H), 3.53 (ddd, J = 11.7, 9.8, 2.6 Hz, 1H), 3.33 – 3.05 (m, 4H), 3.02 – 2.87 (m, 2H), 2.83 – 2.68 (m, 1H), 2.51 (s, 3H).

^{13}C NMR (101 MHz, $CDCl_3$) δ [ppm] = 141.8, 139.9, 139.7, 139.5, 138.6, 137.1, 136.5, 135.9, 133.2, 132.6, 132.2 (2C), 132.1, 129.9, 129.7 (2C), 129.3 (2C), 35.6, 35.3, 35.0, 34.3, 21.3.

IR (ATR) $\tilde{\nu}$ = 2924, 905, 849, 819, 806, 798, 737, 730, 715, 698, 652, 636, 594, 569, 554, 513, 487, 445, 418, 375 cm^{-1}.

HRMS ($C_{23}H_{22}$) calc. 298.1721 , found 298.1723.

The analytical and spectroscopical data match those reported in literature.[197]

4-(4-Methoxyphenyl)[2.2]paracyclophane (74)

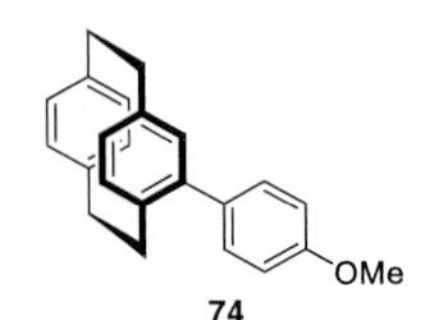

According to general procedure **3.3B**, the title compound was obtained by column chromatography as a white solid (174 mg, 86%).

R$_f$ = 0.25 (pentane/ethyl acetate 50:1).

^{1}H NMR (500 MHz, Chloroform-*d*) δ [ppm] = 7.48 – 7.39 (m, 2H), 7.08 – 7.00 (m, 2H), 6.65 (ddd, J = 7.5, 5.4, 1.9 Hz, 2H), 6.61 – 6.48 (m, 5H), 3.90 (s, 3H), 3.54 – 3.43 (m, 1H), 3.22 – 2.84 (m, 6H), 2.73 – 2.63 (m, 1H).

^{13}C NMR (101 MHz, $CDCl_3$) δ [ppm] = 158.77, 141.52, 139.96, 139.78, 139.50, 137.04, 135.90, 134.10, 133.24, 132.69, 132.11 (2C), 131.92, 130.88 (2C), 129.82, 114.05 (2C), 77.48, 77.16, 76.84, 55.43, 35.66, 35.37, 34.96, 34.36.

IR (ATR) $\tilde{\nu}$ = 3033, 3003, 2925, 2891, 2850, 2834, 1606, 1592, 1574, 1511, 1479, 1460, 1453, 1439, 1418, 1400, 1384, 1290, 1241, 1174, 1108, 1092, 1061, 1041, 1026, 958, 938, 907, 850, 832, 810, 798, 786, 730, 717, 696, 681, 649, 630, 591, 577, 554, 514, 493, 459, 443, 422, 398, 387, 375.

HRMS $C_{23}H_{22}O$ calc. 314.1671, found 314.1669.

The analytical and spectroscopical data match those reported in literature.[197]

4-(4'-Aminophenyl)[2.2]paracyclophane (75)

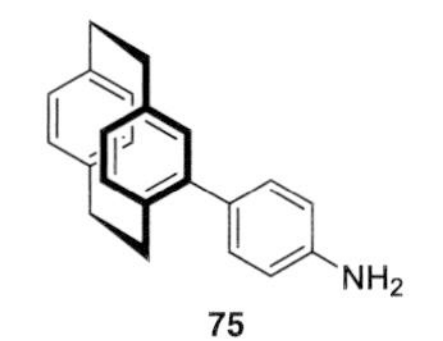

According to general procedure **3.3B**, the title compound was obtained by column chromatography as an off-white solid (44 mg, 23%).

R$_f$ = 0.23 (pentane/ethyl acetate 4:1 + 1% NEt_3).

^{1}H NMR (500 MHz, Chloroform-*d*) δ [ppm] = 7.34 – 7.29 (m, 2H), 6.85 – 6.79 (m, 2H), 6.65 (dt, J = 7.8, 1.8 Hz, 2H), 6.60 – 6.45 (m, 5H), 3.82 (br, 2H), 3.58 – 3.41 (m, 1H), 3.20 – 2.99 (m, 4H), 2.89 (dqt, J = 10.1, 6.9, 3.2 Hz, 2H), 2.77 – 2.65 (m, 1H).

^{13}C NMR (101 MHz, $CDCl_3$) δ [ppm] = 145.0, 141.9, 140.0, 139.7, 139.4, 136.9, 135.8, 133.2, 132.6, 132.3, 132.1, 131.9, 131.6, 130.8 (2C), 129.8, 115.5 (2C), 35.6, 35.4, 34.9, 34.4.

IR (ATR) $\tilde{\nu}$ = 3445 (w), 3355 (w), 2926 (w), 2848 (w), 1622 (w), 1605 (w), 1514 (w), 1478 (w), 1451 (w), 1431 (w), 1401 (w), 1284 (w), 1179 (w), 1128 (vw), 1066 (vw), 940 (vw), 901 (w), 832 (w), 815 (w), 735 (w), 716 (w), 652 (w), 634 (w), 590 (vw), 553 (w), 513 (w), 489 (w), 420 (vw).

HRMS ($C_{22}H_{21}N_1$) calc. 299.1674, found 299.1674.

4-(4'-Diphenylaminophenyl)[2.2]paracyclophane (76)

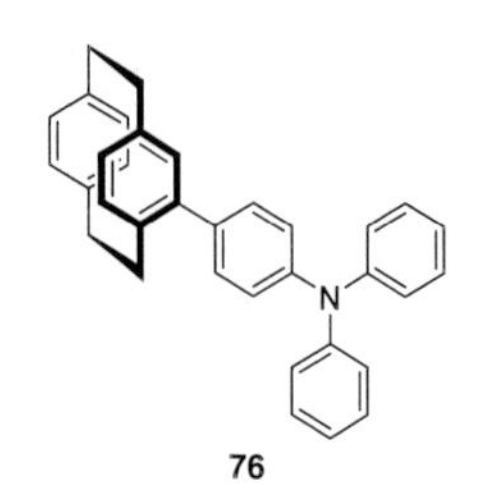

According to general procedure **3.3B**, the title compound was obtained by column chromatography as a yellow solid (220 mg, 88%).

R$_f$ = 0.31 (pentane/ethyl acetate 100:1).

^{1}H NMR (500 MHz, Chloroform-*d*) δ [ppm] = 7.38 – 7.27 (m, 6H), 7.21 – 7.14 (m, 6H), 7.06 (tt, *J* = 7.3, 1.2 Hz, 2H), 6.65 (ddd, *J* = 7.7, 5.7, 1.9 Hz, 2H), 6.60 – 6.47 (m, 5H), 3.57 – 3.46 (m, 1H), 3.22 – 3.08 (m, 3H), 3.07 – 2.98 (m, 1H), 2.96 – 2.85 (m, 2H), 2.81 – 2.70 (m, 1H).

^{13}C NMR (101 MHz, $CDCl_3$) δ [ppm] = 147.9, 146.7, 141.6, 140.0, 139.8, 139.5, 137.1, 135.9, 135.5, 133.3, 132.7, 132.1, 132.1, 132.0, 130.5 (2C), 129.9, 129.4 (4C), 124.7 (4C), 123.5 (2C), 123.1 (2C), 35.7, 35.4, 35.1, 34.5.

IR (ATR) $\tilde{\nu}$ = 2923 (w), 2850 (w), 1587 (m), 1509 (w), 1484 (m), 1314 (w), 1272 (m), 1176 (w), 1074 (w), 1028 (vw), 898 (w), 837 (w), 794 (w), 752 (m), 717 (w), 694 (m), 651 (w), 636 (w), 622 (w), 588 (w), 555 (w), 509 (m), 490 (w).

HRMS ($C_{34}H_{29}N_1$) calc. 451.2300, found 451.2301.

4-(4-(2'-Pyridyl)phenyl)[2.2]paracyclophane (77)

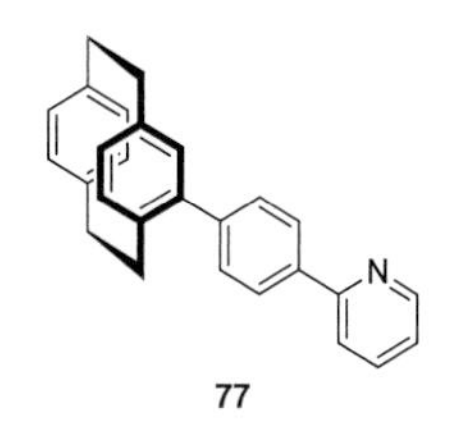

According to general procedure **3.3B**, the title compound was obtained by column chromatography as an off-white solid (127 mg, 90%).

R$_f$ = 0.22 (pentane/ethyl acetate 10:1).

^{1}**H NMR** (500 MHz, Chloroform-*d*) δ [ppm] = 8.75 (dt, *J* = 4.8, 1.4 Hz, 1H), 8.12 (d, *J* = 8.3 Hz, 2H), 7.82 (dt, *J* = 4.7, 1.6 Hz, 2H), 7.68 – 7.53 (m, 2H), 7.30 – 7.27 (m, 1H), 6.69 – 6.49 (m, 7H), 3.50 (ddd, *J* = 12.5, 10.0, 2.9 Hz, 1H), 3.24 – 3.02 (m, 4H), 3.00 – 2.83 (m, 2H), 2.66 (ddd, *J* = 13.0, 9.9, 4.6 Hz, 1H).

^{13}**C NMR** (101 MHz, $CDCl_3$) δ [ppm] = 157.3, 149.7, 142.1, 141.4, 139.9, 139.6, 137.4, 137.2, 136.1, 133.3, 132.7, 132.5, 132.2, 132.2, 130.3 (2C), 130.0, 127.2 (2C), 122.3, 120.8, 35.7, 35.4, 35.0, 34.4.

IR (ATR) $\tilde{\nu}$ = 2923 (m), 2850 (w), 1585 (m), 1572 (m), 1500 (w), 1463 (m), 1432 (m), 1411 (w), 1391 (w), 1294 (w), 1152 (w), 1092 (w), 1060 (vw), 1013 (w), 988 (w), 906 (m), 848 (m), 781 (s), 728 (s), 649 (m), 616 (w), 591 (w), 568 (w), 517 (m), 493 (w), 402 (w).

HRMS ($C_{27}H_{23}N_1$) calc. 361.1830, found 361.1831.

4-(4'-Trimethylsilylethynylphenyl)[2.2]paracyclophane (78)

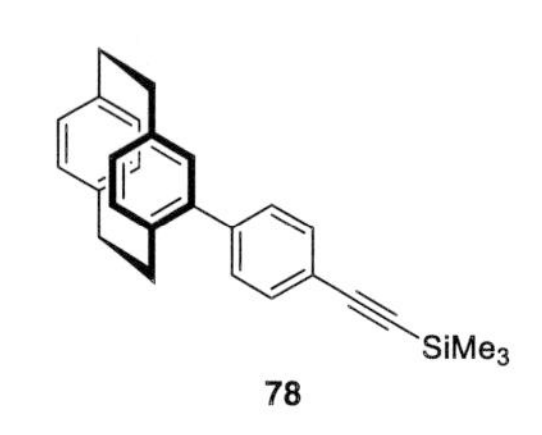

According to general procedure **3.3B**, the title compound was obtained by column chromatography as an off-white solid (156 mg, 78%).

R_f = 0.22 (pentane/ethyl acetate 200:1).

^{1}H NMR (400 MHz, Chloroform-*d*) δ [ppm] = 7.60 (d, J = 8.3 Hz, 2H), 7.44 (d, J = 8.2 Hz, 2H), 6.70 – 6.47 (m, 7H), 3.46 – 3.37 (m, 1H), 3.21 – 3.01 (m, 4H), 2.96 – 2.83 (m, 2H), 2.65 – 2.57 (m, 1H), 0.31 (s, 9H).

^{13}C NMR (101 MHz, $CDCl_3$) δ [ppm] = 141.5, 141.2, 140.0, 139.8, 139.5, 137.2, 136.1, 133.3, 133.1, 132.8, 132.6, 132.3 (2C), 132.1, 132.1, 129.8, 129.7 (2C), 121.5, 105.2, 94.9, 35.6, 35.3, 34.9, 34.3, 0.2.

IR (ATR) $\tilde{\nu}$ = 2952, 2925, 2894, 2851, 2159, 1589, 1503, 1476, 1452, 1436, 1412, 1392, 1259, 1247, 1225, 1180, 1160, 1106, 1092, 1033, 1016, 939, 909, 861, 836, 812, 756, 737, 717, 703, 652, 642, 636, 603, 588, 562, 526, 514, 493, 466, 394 cm^{-1}.

HRMS ($C_{27}H_{28}Si_1$) calc. 380.1960, found 380.1959.

4-(9'-Phenanthrenyl[2.2]paracyclophane (79)

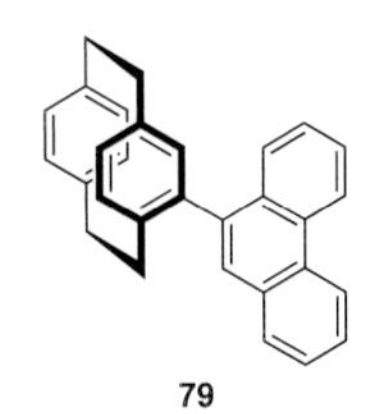

According to general procedure **3.3B**, the title compound was obtained by column chromatography as an off-white solid (26.1 mg, 34%).

R_f = 0.21 (pentane).

^{1}H NMR (400 MHz, Chloroform-*d*) δ [ppm] = 8.77 (d, *J* = 8.1 Hz, 2H), 8.13 (dd, *J* = 7.2, 2.4 Hz, 1H), 8.05 (s, 1H), 7.92 (d, *J* = 8.4 Hz, 1H), 7.72 (ddd, *J* = 5.4, 3.9, 2.1 Hz, 2H), 7.64 (ddd, *J* = 8.4, 6.8, 1.5 Hz, 1H), 7.46 (td, *J* = 7.1, 3.5 Hz, 1H), 6.85 (dd, *J* = 7.8, 2.1 Hz, 1H), 6.80 – 6.70 (m, 3H), 6.68 – 6.61 (m, 3H), 3.28 – 2.58 (m, 8H).

^{13}C NMR (101 MHz, $CDCl_3$) δ [ppm] = 140.2, 140.2, 139.9, 139.8, 138.9, 138.2, 134.9, 133.4, 133.1, 133.0, 132.6, 132.3, 132.1, 131.3, 130.5, 130.1, 129.3, 129.0, 128.0, 126.9, 126.8, 126.7 (2C), 126.6, 122.9, 122.7, 35.7, 35.5, 35.2, 34.5.

IR (ATR) $\tilde{\nu}$ = 2919, 2849, 1735, 1718, 1592, 1492, 1477, 1463, 1449, 1432, 1421, 1411, 1367, 1282, 1259, 1232, 1204, 1179, 1157, 1149, 1133, 1094, 1082, 1035, 1018, 953, 948, 941, 897, 877, 863, 817, 806, 792, 769, 751, 724, 666, 652, 633, 615, 598, 577, 569, 560, 510, 486, 426, 402, 382.

HRMS ($C_{30}H_{24}$) calc. 384.1878, found 384.1877.

4-(3'-Thiophenyl)[2.2]paracyclophane (80)

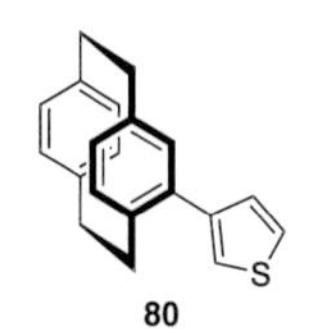

According to general procedure **3.3B**, the title compound was obtained by column chromatography as an off-white solid (190 mg, 87%).

R$_f$ = 0.28 (pentane/ethyl acetate 50:1).

^{1}H NMR (400 MHz, Chloroform-*d*) δ [ppm] = 7.43 (dd, J = 4.9, 2.9 Hz, 1H), 7.34 (dd, J = 2.9, 1.3 Hz, 1H), 7.29 – 7.24 (m, 1H), 6.67 (ddd, J = 15.1, 7.8, 1.9 Hz, 2H), 6.61 – 6.48 (m, 5H), 3.62 – 3.50 (m, 1H), 3.25 – 3.11 (m, 3H), 3.10 – 3.00 (m, 1H), 3.00 – 2.85 (m, 2H), 2.78 – 2.65 (m, 1H).

^{13}C NMR (101 MHz, $CDCl_3$) δ [ppm] = 142.7, 139.9, 139.8, 139.5, 137.4, 136.9, 135.7, 133.3, 132.8, 132.8, 132.1, 132.0, 129.8, 129.0, 125.5, 122.3, 35.6, 35.3, 34.9, 34.5.

IR (ATR) $\tilde{\nu}$ = 2928 (w), 2848 (w), 1590 (vw), 1500 (vw), 1450 (vw), 1357 (vw), 1242 (vw), 1114 (w), 943 (vw), 903 (w), 862 (w), 841 (w), 801 (w), 783 (w), 720 (w), 691 (w), 673 (w), 653 (w), 581 (w), 514 (w), 489 (w), 386 (vw).

HRMS ($C_{20}H_{18}S+H^+$) calc. 291.1202, found 291.1196.

4-(Mesityl)[2.2]paracyclophane (81)

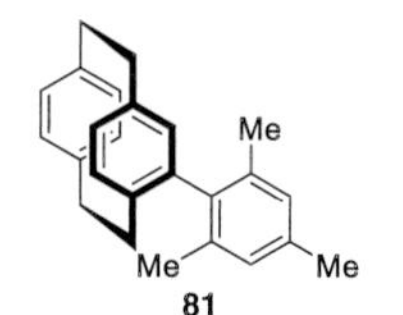

According to general procedure **3.3B**, the title compound was obtained by column chromatography as an off-white solid (62.3 mg, 25%).

R$_f$ = 0.27 (pentane/ethyl acetate 200:1).

^{1}H NMR (400 MHz, Chloroform-*d*) δ [ppm] = 7.09 (d, J = 1.8 Hz, 1H), 6.91 – 6.87 (m, 1H), 6.81 (d, J = 1.2 Hz, 2H), 6.59 (d, J = 1.8 Hz, 1H), 6.55 – 6.49 (m, 1H), 6.48 – 6.41 (m, 2H), 6.34 (dd, J = 7.8, 1.7 Hz, 1H), 3.30 (dddd, J = 15.8, 13.0, 10.3, 2.7 Hz, 2H), 3.16 – 3.02 (m, 3H), 2.99 – 2.88 (m, 2H), 2.85 (s, 3H), 2.78 – 2.66 (m, 1H), 2.36 (s, 3H), 1.85 (s, 3H).

^{13}C NMR (101 MHz, $CDCl_3$) δ [ppm] = 140.1, 139.8, 139.4, 138.5, 137.5, 137.4, 136.4, 136.2, 135.5, 134.7, 133.0, 132.6, 132.4, 132.3, 132.3, 129.9, 129.7, 128.9, 35.5, 35.3, 35.3, 33.1, 22.1, 21.3, 21.0.

IR (ATR) $\tilde{\nu}$ = 2921 (vw), 2849 (vw), 1612 (vw), 1586 (vw), 1470 (vw), 1433 (vw), 1411 (vw), 1091 (vw), 1028 (vw), 939 (vw), 904 (vw), 851 (vw), 840 (vw), 739 (vw), 717 (vw), 645 (vw), 609 (vw), 508 (vw), 483 (vw), 425 (vw).

HRMS ($C_{25}H_{26}$) calc. 326.2035, found 326.2034.

The analytical and spectroscopical data match those reported in literature.[197]

4-(4'-Cyanobenzyl)[2.2]paracyclophane (82)

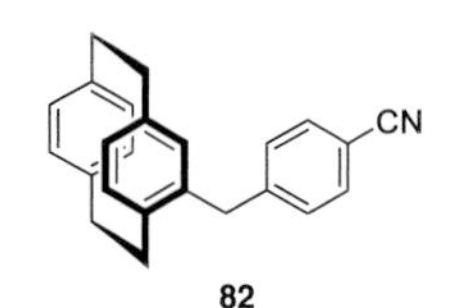

According to general procedure **3.3B**, the title compound was obtained by column chromatography as an off-white solid (66.4 mg, 51%).

R$_f$ = 0.21 (pentane/ethyl acetate 50:1).

^{1}H NMR (500 MHz, Chloroform-*d*) δ [ppm] = 7.51 (d, J = 8.3 Hz, 2H), 7.17 (d, J = 8.2 Hz, 2H), 6.80 (dd, J = 7.9, 1.9 Hz, 1H), 6.60 – 6.40 (m, 5H), 6.05 (d, J = 1.7 Hz, 1H), 4.02 (d, J = 15.8 Hz, 1H), 3.68 (d, J = 15.8 Hz, 1H), 3.32 – 3.13 (m, 2H), 3.13 – 3.01 (m, 4H), 2.98 – 2.90 (m, 1H), 2.81 (ddd, J = 13.5, 10.7, 6.1 Hz, 1H).

^{13}C NMR (101 MHz, $CDCl_3$) δ [ppm] = 146.6, 140.4, 139.7, 139.4, 138.5, 138.0, 135.4, 135.2, 133.6, 133.4, 132.3 (2C), 132.2, 131.4, 129.7 (2C), 128.6, 119.2, 109.9, 40.6, 35.4, 35.1, 34.3, 33.7.

IR (ATR) $\tilde{\nu}$ = 3004, 2951, 2924, 2895, 2850, 2230, 1604, 1591, 1499, 1489, 1453, 1431, 1412, 1261, 1173, 1103, 1086, 1020, 941, 919, 899, 873, 836, 817, 805, 796, 766, 717, 690, 663, 645, 608, 574, 545, 517, 494, 463, 446, 431, 392, 382.

HRMS ($C_{24}H_{22}N_1$) calc. 324.1752, found 324.1754.

4-Bromo-16-(4'-(2''-pyridyl)phenyl)[2.2]paracyclophane (86)

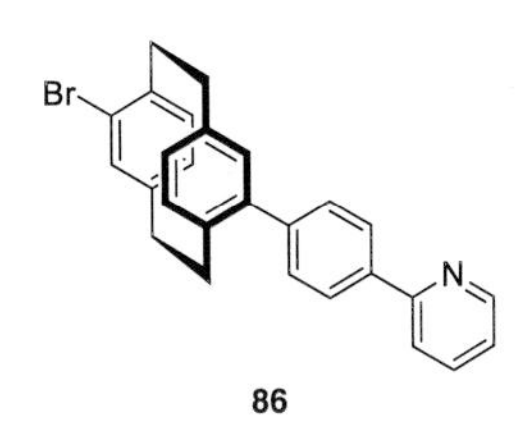

In a vial fitted with a magnetic stirring bar, potassium 4-bromo-16-trifluoroborate[2.2]paracyclophane (1.00 equiv.), potassium phosphate (4.00 equiv.), palladium acetate (0.05 equiv.), RuPhos (0.15 equiv.) and 4-bromo(2'-pyridyl)phenyl (1.00 equiv.) were placed. The vial was evacuated and backfilled with argon three times. After addition of the solvent (Tol:H_2O 1:1), the mixture was put into a vial heating block and heated to 80 °C for 24 h. The reaction was cooled to ambient temperature and quenched with sat. aq. ammonium chloride. After separation of the phases, the aqueous phase was extracted with dichloromethane (3 × 15 mL). The organic phases were dried over sodium sulfate and the solvent was removed under reduced pressure. The crude product was purified *via* column chromatography to afford the title compound as an off-white solid (1.40 g, 23 %.).

R_f = 0.28 (pentane/ethyl acetate 10:1).

^{1}H NMR (500 MHz, Chloroform-*d*) δ [ppm] = 8.76 (dd, J = 4.9, 1.4 Hz, 1H), 8.12 (d, J = 8.4 Hz, 2H), 7.83 (dd, J = 3.4, 1.4 Hz, 2H), 7.59 (d, J = 8.3 Hz, 2H), 7.30 (td, J = 5.0, 3.3 Hz, 1H), 7.18 (dd, J = 7.8, 1.9 Hz, 1H), 6.64 (dd, J = 6.7, 1.6 Hz, 2H), 6.62 – 6.55 (m, 3H), 3.56 (ddd, J = 13.0, 10.3, 2.3 Hz, 1H), 3.45 (ddd, J = 13.7, 9.5, 5.1 Hz, 1H), 3.27 (ddd, J = 13.0, 10.2, 5.4 Hz, 1H), 3.11 – 3.03 (m, 1H), 3.03 – 2.88 (m, 2H), 2.73 (ddd, J = 9.2, 7.1, 4.7 Hz, 2H).

^{13}C NMR (101 MHz, $CDCl_3$) δ [ppm] = 157.0, 149.4, 142.2, 142.0, 141.8, 139.7, 139.0, 137.6, 137.5, 136.9, 135.3, 134.2, 132.2, 130.4, 129.1, 128.8, 127.4, 126.6, 122.4, 120.9, 35.8, 34.5, 33.8, 33.4.

IR (ATR) $\tilde{\nu}$ = 2925 (vw), 1585 (w), 1464 (w), 1433 (w), 1390 (vw), 1186 (vw), 1095 (vw), 1032 (w), 1013 (vw), 907 (vw), 852 (w), 777 (w), 734 (w), 707 (w), 670 (vw), 654 (w), 617 (vw), 572 (vw), 494 (vw), 472 (vw), 398 (vw).

HRMS ($C_{27}H_{22}N_1{}^{79}Br_1$) calc. 439.0936, found 439.0934.

5.2.3 Synthetic Methods and Characterization Data for Chapter 3.4

4-(Trimethylsilylethynyl)-12-formyl[2.2]paracyclophane (104)

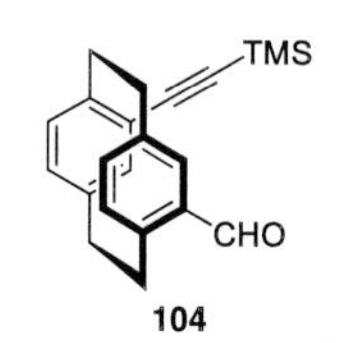

In a Schlenk round bottom flask, 4-bromo-12-formyl[2.2]paracyclophane (500 mg, 1.57 mmol, 1.00 equiv.), trimethylsilylacetylene (1.58 mL, 11.1 mmol, 7.00 equiv.), $PdCl_2(PPh_3)_2$ (111 mg, 0.16 mmol, 10 mol%), triphenylphosphine (83 mg, 0.317 mmol, 20 mol%) and copper(I) iodide (30.2 mg, 0.157 mmol, 10 mol%) were dissolved in THF:NEt_3 (5:2 v/v, 11 mL). The solution was stirred at 75 °C for 2 days under a nitrogen atmosphere. Precipitated ammonium salts were filtered off and the filtrate was evaporated under vacuum. The residue was subjected to column chromatography on silica with pentane/ethyl acetate 50:1 to give the title compound as a yellow oil (322 mg, 61%).

R_f = 0.33 (pentane/ethyl acetate 50:1).
^{1}H NMR (400 MHz, Chloroform-*d*) δ 9.85 (s, 1H), 7.50 (d, *J* = 2.0 Hz, 1H), 6.70 (ddd, *J* = 7.8, 2.0, 0.7 Hz, 1H), 6.65 – 6.54 (m, 2H), 6.47 (d, *J* = 7.9 Hz, 1H), 6.38 (d, *J* = 1.9 Hz, 1H), 4.17 (ddd, *J* = 12.8, 10.0, 1.6 Hz, 1H), 3.62 (ddd, *J* = 12.8, 10.2, 2.1 Hz, 1H), 3.28 (ddd, *J* = 13.1, 10.2, 6.2 Hz, 1H), 3.16 (ddd, *J* = 12.7, 10.4, 1.5 Hz, 2H), 3.00 (ddd, *J* = 12.9, 10.0, 6.8 Hz, 1H), 2.85 (dddd, *J* = 26.9, 13.1, 10.3, 6.5 Hz, 2H), 0.33 (s, 9H). **^{13}C NMR** (101 MHz, $CDCl_3$) δ 192.8, 142.7, 142.6, 140.5, 140.0, 138.5, 136.6, 136.4, 136.1, 135.1, 133.9, 132.8, 125.0, 105.0, 98.9, 34.4, 34.1, 34.0, 33.1, 0.2 (3x).
IR (ATR, ṽ) = 402, 416, 436, 449, 465, 470, 501, 543, 575, 609, 628, 645, 700, 721, 759, 779, 799, 840, 868, 882, 902, 933, 953, 979, 997, 1067, 1098, 1112, 1146, 1159, 1184, 1200, 1224, 1248, 1288, 1322, 1402, 1434, 1451, 1482, 1554, 1588, 1686, 1768, 1795, 1803, 1820, 1837, 1849, 1860, 1871, 1895, 1904, 1921, 1943, 1983, 2013, 2050, 2058, 2064, 2070, 2108, 2146, 2720, 2781, 2854, 2895, 2929, 2956, 3010 cm^{-1}.
HRMS ($C_{22}H_{24}OSi + H^+$) calc. 333.1675; found 333.1675.

4-(Trimethylsilylethynyl)-12-formyl[2.2]paracyclophane (105)

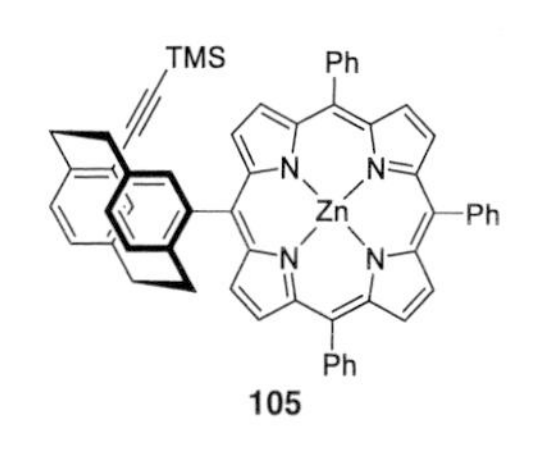

In a 1 L round bottom flask pyrrole (242 mg, 3.61 mmol, 4.00 equiv.) and benzaldehyde (287 mg, 2.70 mmol, 3.00 equiv.) were dissolved in 500 mL dichloromethane and 4-bromo-16-formyl[2.2]paracyclophane (300 mg, 0.920 mmol, 1.00 equiv.) was added. The solution was sparged with argon gas for 15 min. Then, TFA was added (1.21 equiv.). The mixture was stirred for 13.5 h in the dark during which the solution turned dark orange. DDQ (494 mg, 2.18 mmol, 2.54 equiv.) was added and the mixture was stirred for 1 h. After this, the reaction was quenched with triethylamine (5 mL) and the solvents reduced under reduced pressure. The residue was subjected to column chromatography (silica, cyclohexane/dichloromethane 3:1) to afford the title product as a crystalline dark purple solid (23.1 mg, 3%).

R_f = 0.33 (pentane/ethyl acetate 50:1).

^{1}H NMR (400 MHz, Chloroform-*d*) δ 9.85 (s, 1H), 7.50 (d, *J* = 2.0 Hz, 1H), 6.70 (ddd, *J* = 7.8, 2.0, 0.7 Hz, 1H), 6.65 – 6.54 (m, 2H), 6.47 (d, *J* = 7.9 Hz, 1H), 6.38 (d, *J* = 1.9 Hz, 1H), 4.17 (ddd, *J* = 12.8, 10.0, 1.6 Hz, 1H), 3.62 (ddd, *J* = 12.8, 10.2, 2.1 Hz, 1H), 3.28 (ddd, *J* = 13.1, 10.2, 6.2 Hz, 1H), 3.16 (ddd, *J* = 12.7, 10.4, 1.5 Hz, 2H), 3.00 (ddd, *J* = 12.9, 10.0, 6.8 Hz, 1H), 2.85 (dddd, *J* = 26.9, 13.1, 10.3, 6.5 Hz, 2H), 0.33 (s, 9H). **^{13}C NMR** (101 MHz, $CDCl_3$) δ 192.8, 142.7, 142.6, 140.5, 140.0, 138.5, 136.6, 136.4, 136.1, 135.1, 133.9, 132.8, 125.0, 105.0, 98.9, 34.4, 34.1, 34.0, 33.1, 0.2 (3x).

IR (ATR, ṽ) = 402, 416, 436, 449, 465, 470, 501, 543, 575, 609, 628, 645, 700, 721, 759, 779, 799, 840, 868, 882, 902, 933, 953, 979, 997, 1067, 1098, 1112, 1146, 1159, 1184, 1200, 1224, 1248, 1288, 1322, 1402, 1434, 1451, 1482, 1554, 1588, 1686, 1768, 1795, 1803, 1820, 1837, 1849, 1860, 1871, 1895, 1904, 1921, 1943, 1983, 2013, 2050, 2058, 2064, 2070, 2108, 2146, 2720, 2781, 2854, 2895, 2929, 2956, 3010 cm^{-1}.

HRMS ($C_{22}H_{24}OSi + H^+$) calc. 333.1675; found 333.1675.

Triphenylphosphine gold(I) chloride (108)

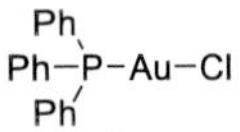

Ph
Ph–P–Au–Cl
Ph
108

In a Schlenk round bottom flask under argon, tetrahydrothiophene gold(I) chlorid and triphenylphosphine were dissolved in degassed dichloromethane. The mixture was stirred in the dark at r.t. for two hours. After this time, diethyl ether was added, and needle-like crystals formed after a few seconds. The solvent was removed by filtration and the residue washed with cold diethyl ether and water to yield the product as a white crystalline solid (96 mg, 78%).

^{1}**H NMR** (500 MHz, Chloroform-*d*) δ 7.58 – 7.42 (m, 1H).

^{31}**P NMR** (202 MHz, Chloroform-*d*) δ 33.20.

The analytical data is in agreement with the data reported in literature.[198]

4-Formyl-16-(4-(2'-pyridyl)phenyl)[2.2]paracyclophane (111)

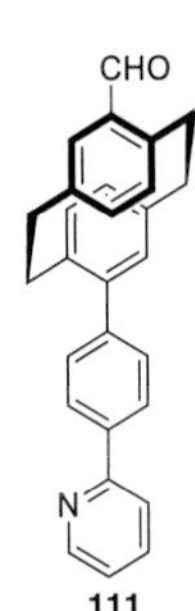

In a vial fitted with a magnetic stirring bar, 4-bromo-16-formyl[2.2]paracyclophane (2.00 g, 6.35 mmol, 1.00 equiv.), potassium phosphate (5.39 g, 373 mmol, 4.00 equiv.), palladium acetate (71.2 mg, 0.317 mmol, 0.05 equiv.), RuPhos (444 mg, 0.952 mmol, 0.15 equiv.) and the phenylpyridine boronic acid (1.89 g, 9.52 mmol, 1.50 equiv.) were placed. The vial was evacuated and backfilled with argon three times. After addition of the solvent (Tol/H_2O 10:1), the mixture was put into a vial heating block and heated to 80 °C overnight. The reaction was cooled to ambient temperature and purified by column chromatography (pentane/ethyl acetate 2:1 to 1:1) to obtain the title product as an off-white powder (260 mg, 11%).

R_f = 0.15 (pentane/ethyl acetate 2:1)

^{1}H NMR (500 MHz, Chloroform-*d*) δ 10.07 (s, 1H), 8.79 (dt, *J* = 4.8, 1.5 Hz, 1H), 8.22 – 8.11 (m, 2H), 7.88 – 7.78 (m, 2H), 7.63 – 7.57 (m, 2H), 7.34 – 7.26 (m, 1H), 7.14 (d, *J* = 1.9 Hz, 1H), 6.85 (dd, *J* = 7.8, 2.0 Hz, 1H), 6.73 (dd, *J* = 4.9, 2.9 Hz, 2H), 6.59 (d, *J* = 7.8 Hz, 1H), 6.45 (dd, *J* = 7.8, 2.0 Hz, 1H), 4.22 (ddd, *J* = 12.6, 10.1, 2.0 Hz, 1H), 3.48 (ddd, *J* = 13.7, 10.0, 4.8 Hz, 1H), 3.29 (ddd, *J* = 12.6, 10.3, 1.9 Hz, 1H), 3.20 – 3.13 (m, 1H), 3.13 – 3.04 (m, 2H), 2.93 (ddd, *J* = 13.4, 10.2, 4.7 Hz, 1H), 2.84 (ddd, *J* = 13.8, 10.0, 4.1 Hz, 1H).

^{13}C NMR (126 MHz, $CDCl_3$) δ 192.1, 157.1, 149.8, 143.3, 141.8, 141.6, 140.9, 139.7, 138.1, 137.2, 136.9, 136.6, 135.2, 135.1, 135.0, 131.8, 130.1, 127.3, 122.3, 120.6, 35.0, 34.5, 33.5, 33.4.

4-Formyl-12-(4-(2'-pyridyl)phenyl)[2.2]paracyclophane (115)

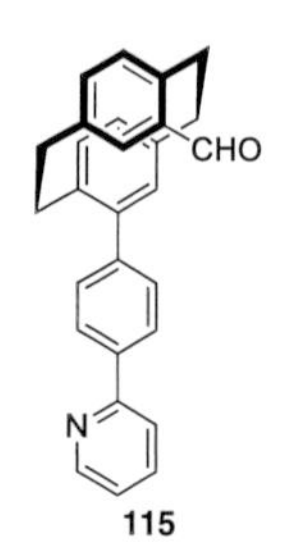

In a vial fitted with a magnetic stirring bar, 4-bromo-12-formyl[2.2]paracyclophane (676 mg, 2.14 mmol, 1.00 equiv.), potassium phosphate (1.82 g, 8.58 mmol, 4.00 equiv.), palladium acetate (48.2 mg, 0.215 mmol, 0.05 equiv.), RuPhos (300 mg, 0.643 mmol, 0.15 equiv.) and the phenylpyridine boronic acid (482 mg, 2.42 mmol, 1.13 equiv.) were placed. The vial was evacuated and backfilled with argon three times. After addition of the solvent (Tol/H_2O 10:1), the mixture was put into a vial heating block and heated to 80 °C overnight. The reaction was cooled to ambient temperature and purified by column chromatography (pentane/ethyl acetate 20:1 to 10:1) to obtain the title product as an off-white powder (148 mg, 18%).

R_f = 0.25 (pentane/ethyl acetate 10:1)

^{1}H NMR (500 MHz, Chloroform-*d*) δ 10.17 (s, 1H), 8.78 – 8.70 (m, 1H), 8.15 – 8.07 (m, 2H), 7.85 – 7.75 (m, 2H), 7.55 – 7.49 (m, 2H), 7.28 – 7.18 (m, 1H), 6.81 (dd, *J* = 7.9, 1.9 Hz, 1H), 6.75 (d, *J* = 7.7 Hz, 1H), 6.69 – 6.58 (m, 2H), 6.48 (d, *J* = 1.8 Hz, 1H), 4.19 – 4.09 (m, 1H), 3.69 (ddd, *J* = 13.6, 9.7, 1.7 Hz, 1H), 3.41 – 3.31 (m, 1H), 3.06 – 2.92 (m, 3H), 2.86 (ddd, *J* = 13.6, 9.9, 7.3 Hz, 1H), 2.52 (ddd, *J* = 13.2, 9.7, 7.3 Hz, 1H).

^{13}C NMR (126 MHz, $CDCl_3$) δ 191.4, 157.3, 149.9, 143.7, 141.2, 141.2, 141.0, 139.7, 138.5, 138.2, 137.1, 137.0, 136.4, 136.1, 136.0, 132.5, 132.0, 131.9, 130.0, 127.5, 122.3, 120.7, 35.3, 35.1, 34.0, 33.7.

5.2.4 Synthetic Methods and Characterization Data for Chapter 3.5

4-(2'-Oxazolinyl)[2.2]paracyclophane (129)

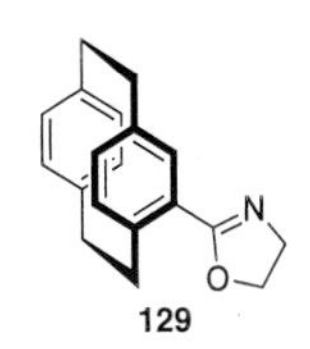

In a flame-dried round bottom flask, 2-aminoethanol (634 µL, 10.6 mmol, 1.00 equiv.) was dissolved in dry dichloromethane (150 mL). In an argon counter-current 4-formyl[2.2]paracyclophane (2.50 g, 10.6 mmol, 1.00 equiv.) and 4 Å molecular sieves (37.5 g) were added. The mixture was stirred slowly for 14 h at r.t. Then, N-bromosuccinimide (1.88 g, 10.6 mmol, 1.00 equiv.) was added and the mixture stirred for 30 minutes at r.t. The yellow suspension was filtered, washed with sat. aq. sodium bicarbonate, washed with water, dried over sodium sulfate and subjected to column chromatography to afford the title compound as a white solid (1.68 g, 57%).

R_f = 0.29 (pentane/ethyl acetate 4:1, 1% NEt_3).
^{1}H NMR (400 MHz, Chloroform-*d*) δ 7.1 (d, *J* = 1.9 Hz, 1H), 6.6 – 6.4 (m, 6H), 4.5 – 4.3 (m, 2H), 4.2 - 4.0 (m, 3H), 3.2 - 2.9 (m, 6H), 2.9 (ddd, *J* = 13.0, 10.1, 6.7 Hz, 1H).
^{13}C NMR (101 MHz, $CDCl_3$) δ 165.0, 141.0, 140.0, 139.6, 139.4, 135.9, 135.0, 134.5, 133.0, 132.8, 132.5, 131.4, 128.4, 66.9, 55.4, 36.0, 35.4, 35.2, 35.0.
IR (ATR) $\tilde{\nu}$ = 2924 (w), 1686 (vw), 1634 (m), 1590 (w), 1478 (w), 1409 (vw), 1348 (w), 1255 (w), 1169 (w), 1126 (w), 1062 (vw), 1040 (m), 892 (w), 946 (w), 907 (w), 887 (w), 824 (w), 740 (w), 709 (w), 631 (m), 514 (m), 385 (vw).
HRMS ($C_{11}9H_{19}O_1N_1$) calc. 277.1467, found 277.1467.

4-Bromo-16-(2'-oxazolinyl)[2.2]paracyclophane (131)

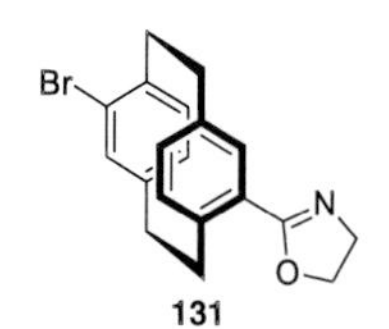

In a flame-dried round bottom flask, 2-aminoethanol (380 µL, 6.34 mmol, 2.00 equiv.) was dissolved in dry dichloromethane (60 mL). In an argon counter-current 4-bromo-16-formyl[2.2]paracyclophane (1.00 g, 3.17 mmol, 1.00 equiv.) and 4 Å molecular sieves (15.0 g) were added. The mixture was stirred slowly for 14 h at r.t. Then, N-bromosuccinimide (1.13 g, 6.34 mmol, 2.00 equiv.) was added and the mixture stirred for 30 minutes at r.t. The yellow suspension was filtered, washed with sat. aq. sodium bicarbonate, washed with water, dried over sodium sulfate and subjected to column chromatography to afford the title compound as a white solid (848 mg, 73%).

R_f = 0.30 (pentane/ethyl acetate 3:1, 1% NEt_3).

^{1}H NMR (400 MHz, Chloroform-*d*) δ 7.24 (dd, *J* = 7.8, 2.0 Hz, 1H), 7.03 (d, *J* = 1.9 Hz, 1H), 6.60 – 6.54 (m, 2H), 6.48 (dd, *J* = 17.6, 7.7 Hz, 2H), 4.52 – 4.32 (m, 2H), 4.20 – 4.07 (m, 3H), 3.46 (ddd, *J* = 13.2, 10.3, 2.7 Hz, 1H), 3.19 (ddd, *J* = 13.1, 10.3, 5.1 Hz, 1H), 3.08 – 2.96 (m, 3H), 2.87 (dddd, *J* = 14.3, 9.1, 6.0, 3.6 Hz, 2H).

^{13}C NMR (101 MHz, $CDCl_3$) δ 165.0, 142.0, 140.6, 139.2, 138.9, 137.0, 135.1, 134.8, 134.4, 131.2, 130.2, 128.5, 127.1, 67.0, 55.4, 35.4, 35.3, 34.2, 33.3.

IR (ATR, ṽ) = 3043, 2963, 2945, 2931, 2917, 2880, 2856, 1635, 1256, 1045, 1034, 986, 949, 939, 909, 899, 887, 836, 826, 704, 690, 664, 645, 523 cm^{-1}.

HRMS ($C_{19}H_{19}O_1N_1{}^{79}Br_1$) calc. 356.0650; found 356.0650.

Tetrahydrothiophene gold(I) chloride (107)

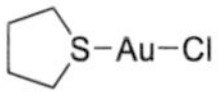

S−Au−Cl

107

In a round bottom flask, $NaAuCl_4$ (800 mg, 2.01 mmol, 1.00 equiv.) was dissolved in water (8.0 mL) and ethanol (19.1 mL). Dropwise addition of tetrahydrothiophene (0.39 mL, 4.42 mmol, 2.20 equiv.) led to a precipitation of a white solid in seconds. The mixture was stirred for 15 minutes during which all color disappeared. After filtration and washing with water and diethyl ether, the product was obtained as a white solid and dried in the dark under reduced pressure (537 mg, 83%).

R_f = 0.30 (pentane/ethyl acetate 3:1, 1% NEt_3).

^{1}H NMR (400 MHz, Chloroform-*d*) δ 3.30 (m, 4H), 2.14 (m, 4H).

^{13}C NMR (101 MHz, $CDCl_3$) δ 38.8, 30.8.

The analytical data is in agreement with the data reported in literature.[199]

4-(Diphenylphosphinyl gold(I) chloride)-16-(2′-oxazolinyl)[2.2]paracyclophane (132)

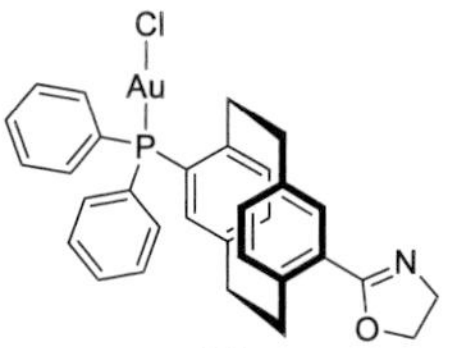

Cl
Au
P
N
O
132

In a flame-dried Schlenk tube, 4-bromo-16-(2'-oxazolinyl[2.2]paracyclophane (115 mg, 323 µmol, 1.00 equiv.) was dissolved in dry and degassed THF (0.05 M). A solution of *n*-BuLi in *n*-hexanes (136 µL, 339 µmol, 1.05 equiv.) was added dropwise at –78 °C. After 30 minutes, chlorodiphenylphosphine (65 µL, 355 µmol, 1.10 equiv.) was added dropwise. The reaction was slowly allowed to warm up to r.t. overnight. In an argon counter-current (tht)gold(I)chloride (114 mg, 355 µmol, 1.00 equiv.) was added to the mixture and stirred for 2 hours at r.t. After removal of the solvent the crude residue was purified by column chromatography to afford the product as a white solid (180 mg, 80%).

R_f = 0.25 (pentane/ethyl acetate 2:1).

^{1}H NMR (500 MHz, $CDCl_3$) δ 7.77 (dt, J = 8.1, 1.4 Hz, 1H), 7.61 – 7.51 (m, 4H), 7.51 – 7.43 (m, 4H), 7.37 (td, J = 7.7, 2.8 Hz, 2H), 7.17 (d, J = 2.0 Hz, 1H), 6.73 (dt, J = 7.7, 1.5 Hz, 1H), 6.56 (dd, J = 7.7, 5.6 Hz, 1H), 6.28 (d, J = 8.0 Hz, 1H), 5.95 (dd, J = 14.8, 1.9 Hz, 1H), 4.58 – 4.34 (m, 2H), 4.17 – 4.08 (m, 2H), 4.02 (t, J = 10.2 Hz, 1H), 3.72 (pt, J = 11.0, 2.0 Hz, 2H), 3.12 – 2.87 (m, 3H), 2.83 – 2.71 (m, 2H).

^{13}C NMR (126 MHz, $CDCl_3$) δ 165.3, 144.4, 144.4, 141.0, 140.9, 140.8, 139.5, 136.1, 136.1, 135.8, 135.7, 135.4, 135.2, 135.1, 135.1, 134.9, 134.5, 134.2, 134.1, 134.0, 132.3, 132.2, 132.1, 132.0, 130.6, 130.1, 129.5, 129.4, 129.2, 129.1, 128.0, 126.9, 126.4, 125.1, 67.3, 55.1, 34.8, 34.7, 34.6, 34.3, 34.3.

^{31}P NMR (202 MHz, $CDCl_3$) δ [ppm] 29.9.

IR (ATR, ṽ) = 3367 (w), 2925 (vw), 1633 (w), 1478 (vw), 1434 (w), 1345 (vw), 1311 (vw), 1259 (vw), 1187 (vw), 1099 (w), 1065 (w), 1043 (vw), 998 (vw), 978 (w), 948 (vw), 919 (vw), 836 (vw), 750 (w), 693 (w), 596 (w), 551 (w), 522 (w), 509 (w), 493 (w), 427 (w).

HRMS ($C_{31}H_{28}AuClNOP + H^+$) calc. 694.1335; found 694.1318.

(Acetonitrile-κ*N*)-(η⁶-p-cymene)-[4-(diphenylphosphinyl gold(I) chloride)-16-(2′-oxazolinyl-κ*N*)[2.2]paracyclophanido-κC^{15}]ruthenium(II) hexafluorophosphate (130)

In a flame-dried Schlenk tube under argon, 4-(diphenylphosphinyl gold(I) chloride)-16-(2′-oxazolinyl)[2.2]paracyclophane (**132**) (64.6 mg, 93 μmol, 1.00 equiv.), $[RuCl(p\text{-cymene})]_2$ (28.5 mg, 47 μmol, 0.50 equiv.), potassium acetate (13.7 mg, 140 μmol, 1.50 equiv.), and potassium hexafluorophosphate (34.3 mg, 186 μmol, 2.00 equiv.) were dissolved in dry acetonitrile (0.02 M). The mixture was degassed by three freeze-pump-thaw cycles and then stirred at r.t. for 3 days. The solvent was removed under reduced pressure and the crude green mixture purified by column chromatography on silica and the yellow band isolated to afford the title compound as a canary yellow solid (79 mg, 88%).

R_f = 0.41 (dichloromethane/acetonitrile 10:1).

^{1}H NMR (400 MHz, $CDCl_3$) δ 7.68 – 7.26 (m, 10H), 6.18 – 6.06 (m, 2H), 6.07 – 5.99 (m, 1H), 5.87 (dd, J = 7.7, 2.0 Hz, 1H), 5.74 (d, J = 6.1 Hz, 1H), 5.63 (dd, J = 6.0, 1.8 Hz, 1H), 5.26 (dd, J = 6.0, 1.9 Hz, 1H), 5.11 – 5.04 (m, 1H), 4.82 (dtt, J = 10.3, 8.2, 2.0 Hz, 1H), 4.75 – 4.65 (m, 1H), 4.13 – 3.91 (m, 3H), 3.67 (t, J = 11.5 Hz, 1H), 3.61 – 3.52 (m, 1H), 3.40 (dt, J = 15.9, 8.4 Hz, 1H), 2.97 (dt, J = 14.5, 8.3 Hz, 1H), 2.84 – 2.74 (m, 1H), 2.70 (t, J = 7.2 Hz, 2H), 2.58 – 2.45 (m, 1H), 1.99 (hept, 1H), 1.84 (s, 3H), 1.71 (s, 3H), 0.77 (dt, J = 7.0, 1.9 Hz, 3H), 0.59 (dt, J = 7.0, 1.8 Hz, 3H).

^{13}C NMR (101 MHz, $CDCl_3$) δ 179.7, 175.8, 150.1, 146.0, 145.9, 141.1, 141.0, 138.8, 136.2, 136.1, 135.8, 135.7, 135.3, 135.2, 134.9, 134.7, 133.3, 133.1, 132.3, 132.2, 131.6, 130.6, 130.4, 130.3, 130.3, 130.1, 106.7, 104.3, 92.3, 91.5, 84.9, 83.7, 72.5, 55.1, 39.4, 35.6, 35.5, 35.1, 32.6, 31.6, 22.3, 21.8, 18.4.

^{31}P NMR (162 MHz, $CDCl_3$) δ 35.3, –139.2 (hept, *J* = 706.4 Hz).

HRMS ($C_{43}H_{44}AuClN_2OPRu$) calc. 969.1589; found 969.1573.

5.2.5 Synthetic Methods and Characterization Data for Chapter 3.6

4-Diphenylphosphoryl-16-(4-(2'-pyridyl)phenyl)[2.2]paracyclophane (163)

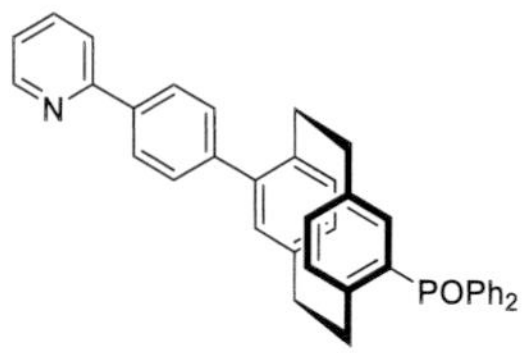

In a vial with a magnetic stirring bar, 4-bromo-16-diphenylphosphoryl[2.2]paracyclophane (100 mg, 0.205 mmol, 1.00 equiv.), 4-bromo-(2'-pyridyl)-phenyl (61.3 mg, 0.308 mmol, 1.50 equiv.), palladium(II) acetate (2.3 mg, 10.2 µmol, 5 mol%), RuPhos (14.3 mg, 30.8 µmol, 15 mol%) and potassium phosphate (174 mg, 0.820 mmol, 4.00 equiv.) were placed. The vial was capped and evacuated and backfilled with argon three times. Then, toluene (2 mL) and water (0.2 mL) were added with a syringe. The mixture was stirred and heated to 80 °C overnight. The next day, the mixture was passed over a Celite pad and the solvent removed under reduced pressure. The crude mixture was subjected to column chromatography (silica, pentane/ethyl acetate 2:1 to 1:1) to yield the title compound as an off-white solid (88 mg, 93%).

R_f = 0.22 (pentane/ethyl acetate). **m.p.:** 265 °C.

^{1}H NMR (400 MHz, $CDCl_3$) δ 8.7 (dt, *J* = 4.8, 1.4 Hz, 1H), 8.1 – 8.0 (m, 2H), 7.8 (dt, *J* = 4.4, 1.6 Hz, 2H), 7.8 – 7.7 (m, 2H), 7.6 – 7.5 (m, 5H), 7.5 (tq, *J* = 6.7, 1.7 Hz, 3H), 7.4 (ddd, *J* = 8.4, 6.7, 2.9 Hz, 2H), 7.3 – 7.2 (m, 2H), 6.7 (dd, *J* = 8.8, 1.6 Hz, 2H), 6.7 (dd, *J* = 7.8, 4.1 Hz, 1H), 6.4 (d, *J* = 7.8 Hz, 1H), 6.4 (dd, *J* = 14.5, 1.8 Hz, 1H), 3.7 – 3.5 (m, 2H), 3.3 (ddd, *J* = 13.6, 10.1, 5.1 Hz, 1H), 3.1 – 2.9 (m, 3H), 2.8 (ddd, *J* = 13.6, 10.0, 3.6 Hz, 1H), 2.6 (ddd, *J* = 13.4, 10.2, 5.1 Hz, 1H).

^{13}C NMR (101 MHz, $CDCl_3$) δ 157.4, 149.9, 146.1, 146.1, 142.1, 141.2, 140.3, 140.0, 139.8, 138.0, 137.3, 137.2, 136.9, 136.7, 135.9, 135.4, 135.3, 135.2, 134.9, 134.1, 133.9, 133.9, 133.4, 132.4, 132.3, 132.2, 131.6, 131.6, 131.6, 131.5, 131.1, 130.3, 130.1, 128.6, 128.5, 128.4, 128.4, 127.3, 122.2, 120.7, 35.6, 35.6, 35.4, 34.9, 33.2.

^{31}P NMR (162 MHz, $CDCl_3$) δ 27.0.

IR (ATR, ṽ) = 3053, 3024, 3006, 2968, 2924, 2890, 2846, 1584, 1572, 1459, 1434, 1394, 1179, 1160, 1116, 1103, 1067, 853, 846, 788, 768, 747, 731, 717, 694, 654, 643, 615, 557, 537, 524, 507, 496, 490, 443, 398 cm^{-1}.

HRMS ($C_{39}H_{33}ONP$) calc. 562.2300; found 562.2299.

4-(Diphenylphosphinyl)-16-(4-(2'-pyridyl)phenyl)[2.2]paracyclophane (164)

= GP phosphine oxide reduction

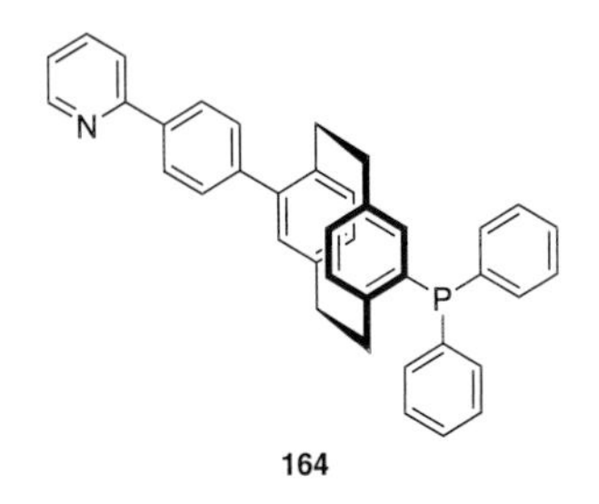

Triethylamine (0.35 mL, 2.50 mmol, 2.00 equiv.) was added dropwise to a stirred solution of 4-diphenylphosphoryl-16-(4-(2'-pyridyl)phenyl)-[2.2]paracyclophane (**163**) (700 mg, 1.00 mmol, 1.00 equiv.) in toluene (7.0 mL) under argon. To this solution was added trichlorosilane (0.25 mL, 2.50 mmol, 2.00 equiv.) and stirring was continued at 120 °C overnight. The reaction was carefully quenched with sat. aq. ammonium chloride and extractive workup was conducted with ethyl acetate. The organic extracts were washed with brine and dried over anhydrous sodium sulfate and used without further purification in the next step. The conversion was determined to be 83% by comparison of the peaks of product and starting material in the ^{31}P NMR spectrum.

^{1}H NMR (400 MHz, Chloroform-*d*) δ 8.79 – 8.68 (m, 1H), 8.09 (d, *J* = 7.8 Hz, 2H), 7.78 (d, *J* = 5.4 Hz, 2H), 7.58 (d, *J* = 7.7 Hz, 2H), 7.41 (s, 4H), 7.25 (dd, *J* = 6.1, 3.5 Hz, 6H), 7.17 (d, *J* = 7.7 Hz, 1H), 6.70 (d, *J* = 2.0 Hz, 1H), 6.66 – 6.52 (m, 2H), 6.32 (d, *J* = 7.9 Hz, 1H), 5.99 – 5.70 (m, 1H), 3.71 – 3.44 (m, 2H), 3.34 (ddd, *J* = 14.1, 10.0, 4.6 Hz, 1H), 3.13 – 3.02 (m, 1H), 2.91 (dp, *J* = 13.7, 6.5, 4.3 Hz, 1H), 2.80 (ddd, *J* = 13.7, 10.2, 4.1 Hz, 1H), 2.63 (ddd, *J* = 13.9, 10.0, 4.2 Hz, 1H), 2.52 (ddd, *J* = 13.8, 10.2, 4.4 Hz, 1H).

^{31}P NMR (162 MHz, Chloroform-*d*) δ -3.53.

4-(Diphenylphosphinyl gold(I) chloride)-16-(4-(2'-pyridyl)phenyl)[2.2]para-cyclophane (165)

=GP phosphine auration

In a round bottom flask, crude 4-(diphenylphosphinyl)-16-(4-(2'-pyridyl)phenyl)-[2.2]paracyclophane (**164**) (420 mg, 0.770 mmol, 1.00 equiv.) and (tht)AuCl (381 mg, 0.770 mmol, 1.00 equiv.) were placed. The round bottom flask was closed with a rubber septum and evacuated and backfilled with argon three times. Degassed DCM (7 mL) was added with a syringe. The mixture was stirred for 2 hours at r.t. in the dark. The solvent was removed under reduced pressure and the residue subjected to column chromatography (silica, pentane/ethyl acetate 2:1) to yield the product as a white solid (277 mg, 46%).

165

R_f = 0.44 (pentane/ethyl acetate 2:1.

^{1}H NMR (500 MHz, Chloroform-*d*) δ 8.7 (dt, *J* = 4.7, 1.4 Hz, 1H), 8.2 – 8.1 (m, 2H), 7.9 – 7.8 (m, 2H), 7.6 – 7.6 (m, 5H), 7.6 – 7.5 (m, 2H), 7.5 – 7.5 (m, 4H), 7.4 (td, *J* = 7.7, 2.7 Hz, 2H), 7.3 – 7.3 (m, 1H), 6.8 (d, *J* = 1.9 Hz, 1H), 6.7 (dt, *J* = 7.7, 1.5 Hz, 1H), 6.6 (dd, *J* = 7.8, 5.5 Hz, 1H), 6.3 (d, *J* = 7.9 Hz, 1H), 6.0 (dd, *J* = 14.7, 1.7 Hz, 1H), 3.9 – 3.8 (m, 2H), 3.3 (ddd, *J* = 13.8, 10.2, 5.0 Hz, 1H), 3.1 (ddd, *J* = 12.4, 11.0, 2.3 Hz, 1H), 3.0 (dddd, *J* = 17.3, 13.9, 10.3, 3.4 Hz, 2H), 2.8 (ddd, *J* = 13.6, 10.1, 3.6 Hz, 1H), 2.6 (ddd, *J* = 13.5, 10.3, 5.0 Hz, 1H).

^{13}C NMR (101 MHz, $CDCl_3$) δ 157.1, 149.7, 144.4, 144.3, 141.8, 141.6, 141.0, 140.9, 139.6, 138.0, 137.2, 137.0, 136.6, 136.6, 135.6, 135.5, 135.4, 135.3, 135.0, 134.2, 134.1, 133.8, 133.8, 132.2, 132.2, 132.0, 132.0, 131.4, 130.8, 130.2, 129.5, 129.4, 129.3, 129.2, 129.1, 128.8, 127.4, 126.8, 126.3, 122.4, 120.8, 34.9, 34.9, 34.7, 34.6, 33.1, 30.6.

^{31}P NMR (162 MHz, $CDCl_3$) δ 29.8 (d, *J* = 14.3 Hz).

IR (ATR, ṽ) = 3050, 3003, 2921, 2851, 1732, 1608, 1584, 1572, 1463, 1434, 1412, 1391, 1309, 1296, 1238, 1183, 1153, 1098, 1011, 993, 911, 849, 782, 742, 710, 691, 652, 616, 571, 548, 523, 510, 490, 443, 426, 401 cm^{-1}.

HRMS ($C_{39}H_{32}AuClNP + H^+$) calc. 778.1699; found 778.1685.

cis-Bis(2,2′-bipyridine)dichlororuthenium(II) hydrate (168)

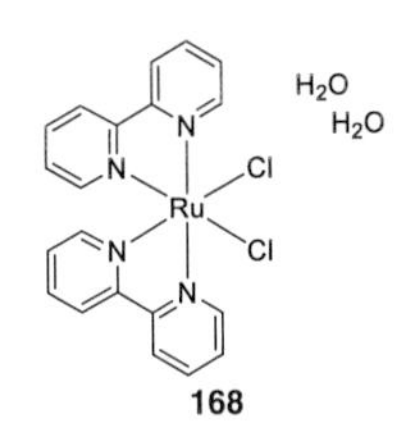

In a 100 mL round bottom flask $RuCl_3 \cdot H_2O$ (5.13 g, 19.6 mmol), 2,2'-bipyridine (6.13 g, 39.2 mmol) and LiCl (5.54 g, 131 mmol) were refluxed and stirred in reagent grade dimethylformamide (65 mL) under an argon atmosphere for 8 h. After the reaction mixture was cooled down to room temperature, acetone (325 mL) was added and the resultant solution was kept at 4 °C overnight. Filtering yielded a dark green-black microcrystalline product. The solid was washed with water (3×25 mL) and diethyl ether (3×25 mL) and then dried in vacuo to give $Ru(bpy)_2Cl_2.2H_2O$ as a green-black powder (5.92 g, 58%).

^{1}H NMR (400 MHz, DMSO-d_6) δ 10.0 (dd, *J* = 6.0, 1.3 Hz, 1H), 8.7 (d, *J* = 8.2 Hz, 1H), 8.6 – 8.6 (m, 1H), 8.5 (dd, *J* = 8.1, 1.3 Hz, 1H), 8.4 (td, *J* = 7.9, 1.6 Hz, 1H), 8.2 (td, *J* = 7.8, 1.5 Hz, 1H), 8.1 – 8.0 (m, 4H), 8.0 (ddd, *J* = 7.3, 5.6, 1.3 Hz, 1H), 7.9 (ddd, *J* = 7.4, 5.7, 1.4 Hz, 1H), 7.8 – 7.8 (m, 1H), 7.8 (ddd, *J* = 7.2, 5.7, 1.3 Hz, 1H), 7.7 (t, *J* = 7.8 Hz, 1H), 7.5 (d, *J* = 5.7 Hz, 1H), 7.4 (dddd, *J* = 16.0, 7.3, 5.6, 1.3 Hz, 3H), 7.1 (dt, *J* = 5.5, 1.2 Hz, 1H), 7.1 (t, *J* = 6.4 Hz, 1H).

^{13}C NMR (101 MHz, DMSO-d_6) δ 140.7, 138.7, 138.2, 137.9, 137.7, 136.4, 135.7, 133.6, 133.6, 133.2, 132.5, 129.5, 119.1, 119.0, 118.2, 117.9, 115.0, 113.8, 108.4, 107.8, 107.4, 105.8, 105.7, 105.2, 104.8, 104.3, 103.9, 103.3, 103.0.

IR (ATR, $\tilde{\nu}$) = 3483, 3466, 3455, 3448, 3437, 3424, 3414, 1612, 1599, 1456, 1441, 1417, 1306, 1259, 1241, 1156, 1122, 1017, 1007, 992, 976, 962, 902, 885, 762, 727, 656, 647 cm^{-1}.

HRMS ($C_{20}H_{16}N_4Cl_2$) calc. 483.9795; found 483.9796.

The analytical data is in agreement with the data reported in literature.[174]

cis-Bis(2,2'-bipyridine)bis(pyridine)ruthenium(II) chloride (171)

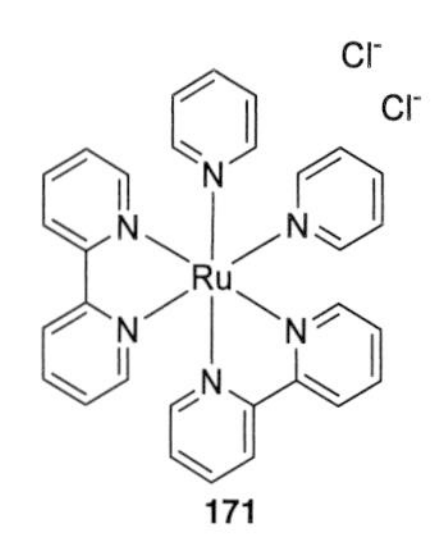

A round bottomed flask was charged with pyridine (8.0 mL), water (12 mL) cm3) and cis-Bis(2,2'-bipyridine)dichlororuthenium(II) hydrate (**121**) (455 mg, 0.874 mmol, 1.00 equiv.). The reaction mixture was stirred while heating at reflux for 4 h, filtered while hot, and the solvent evaporated under reduced pressure. The deep red residue was dissolved in methanol (10 mL) and sufficient diethyl ether (10 mL) was added to result in the formation of a red precipitate. The mixture was allowed to stand at room temperature for 1 h, after which the precipitate was recovered by suction filtration and the crystals were washed with diethyl ether (2 × 50 mL) to afford the product as a red crystalline solid (526 mg, 94%).

^{1}H NMR (400 MHz, Methanol-d_4) δ 9.2 – 9.0 (m, 1H), 8.6 – 8.6 (m, 1H), 8.6 – 8.5 (m, 3H), 8.2 (td, J = 7.9, 1.4 Hz, 1H), 8.1 – 8.0 (m, 2H), 7.9 (tdd, J = 7.2, 5.9, 1.4 Hz, 2H), 7.5 (ddd, J = 7.3, 5.7, 1.3 Hz, 1H), 7.4 – 7.4 (m, 2H).

^{13}C NMR (101 MHz, Methanol-d_4) δ 228.3, 228.1, 224.1, 223.0, 222.9, 208.6, 208.6, 208.2, 198.6, 198.2, 196.8, 194.4, 194.3.

The analytical data is in agreement with the data reported in literature.[200]

4-(((Bis(2,2'-bipyridyl)4-(2'-pyridyl)phenyl) ruthenium (II))-16-(chloro-(diphenylphosphinyl)gold(I))[2.2]paracyclophane (145)

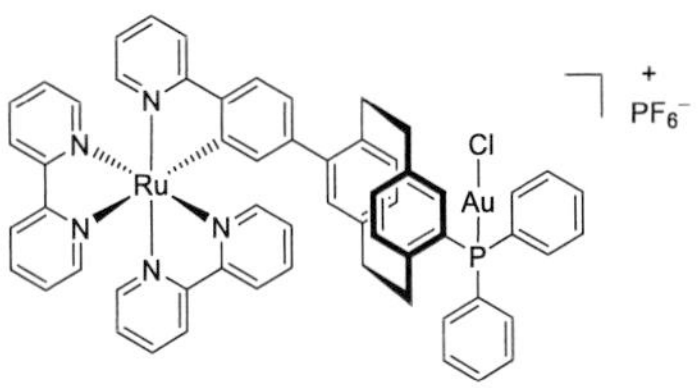

In a vial under argon 4-(4-(2'-pyridyl)-phenyl))16-(chloro(diphenylphosphinyl)-gold(I))[2.2]paracyclophane (30.0 mg, 38.6 µmol, 1.00 equiv.), (Benzene)ruthenium dichloride dimer (9.6 mg, 19.3 µmol, 0.500 equiv.), sodium hydroxide (1.5 mg, 38.6 µmol, 1.00 equiv.) and potassium hexafluorophosphate (14.2 mg, 77.1 µmol, 2.00 equiv.) were dissolved in degassed acetonitrile (2.0 mL) and stirred at 50 °C overnight. The next day, the solvent was removed under reduced pressure. The crude was purified on silica gel with dichloromethane/acetonitrile 10:1 as eluent and the yellow band was collected. This intermediate and 2,2'-bipyridine (12.0 mg, 0.771 mmol, 2.00 equiv.) were dissolved in degassed methanol (4.0 mL) under argon and refluxed overnight. The solution changed from canary yellow to berry-red. The solvent was removed, and the crude purified on silica gel with dichloromethane/acetonitrile 10:1. The first red band was collected to afford the title compound as a very dark red solid (28.0 mg, 76%).

R$_f$ = 0.28 (dichloromethane/acetonitrile 10:1).

^{1}H NMR (400 MHz, Chloroform-*d*) δ 8.72 (dd, *J* = 4.7, 1.5 Hz, 1H), 8.55 (dd, *J* = 18.4, 8.0 Hz, 1H), 8.41 (q, *J* = 7.7 Hz, 1H), 8.24 (d, *J* = 8.0 Hz, 1H), 8.10 (d, *J* = 8.3 Hz, 2H), 8.06 – 7.89 (m, 3H), 7.85 – 7.76 (m, 2H), 7.67 – 7.43 (m, 12H), 7.39 (td, *J* = 7.6, 3.9 Hz, 2H), 7.21 (d, *J* = 6.9 Hz, 1H), 7.04 (d, *J* = 7.7 Hz, 2H), 6.76 (d, *J* = 1.9 Hz, 1H), 6.71 (d, *J* = 7.8 Hz, 1H), 6.63 (dd, *J* = 7.8, 5.4 Hz, 1H), 6.57 (t, *J* = 2.2 Hz, 1H), 6.34 (d, *J* = 7.9 Hz, 1H), 6.29 – 6.19 (m, 2H), 6.11 (dd, *J* = 7.9, 4.6 Hz, 1H), 6.00 (dd, *J* = 14.7, 1.7 Hz, 1H), 5.95 – 5.86 (m, 2H), 3.92 – 3.76 (m, 1H), 3.76 – 3.62 (m, 1H), 3.33 (ddd, *J* = 14.3, 10.2, 5.0 Hz, 1H), 3.11 – 2.70 (m, 4H), 2.60 – 2.52 (m, 1H).

^{13}C NMR (101 MHz, $CDCl_3$) δ 157.2, 156.6, 156.0, 155.8, 150.1, 149.9, 149.8, 144.4, 144.3, 144.0, 143.9, 141.7, 141.6, 141.0, 140.9, 140.9, 140.8, 140.7, 139.6, 139.4, 139.3, 138.2, 137.4, 137.2, 137.1, 137.0, 136.6, 136.5, 136.2, 135.7, 135.6, 135.5, 135.5, 135.4, 135.3, 135.2, 135.2, 135.1, 135.1, 135.0, 134.2, 134.1, 134.0, 133.8, 133.8, 132.3, 132.2, 132.1, 132.0, 131.3, 130.9, 130.8, 130.7, 130.5, 130.2, 130.1, 130.0, 129.5, 129.4, 129.3, 129.3,

129.2, 129.2, 129.1, 128.8, 128.7, 128.4, 127.4, 126.8, 126.6, 126.6, 126.3, 126.2, 126.2, 126.1, 126.0, 125.9, 123.1, 123.0, 122.9, 122.8, 122.7, 122.3, 121.3, 120.7, 34.9, 34.9, 34.8, 34.7, 34.6, 34.5, 34.2, 33.7, 33.1.

IR (ATR, ṽ) = 2917, 2849, 1737, 1715, 1462, 1438, 1421, 1375, 1258, 1239, 1183, 1098, 1041, 1037, 1026, 841, 805, 764, 728, 720, 694, 557 cm^{-1}.

HRMS ($C_{59}H_{47}AuClN_5PRu$) calc. 1190.1961; found 1190.1958.

4-Diphenylphosphine gold(I) chloride[2.2]paracyclophane (191)

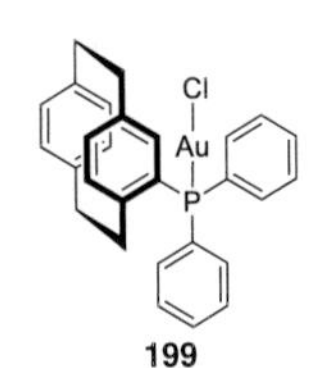

In a flame-dried Schlenk tube, 4-bromo[2.2]paracyclophane (100 mg, 348 µmol, 1.00 equiv.) was dissolved in dry and degassed THF (0.07 M). A solution of *n*-BuLi in *n*-hexanes (139 µL, 348 µmol, 1.00 equiv.) was added dropwise at −78 °C. After 30 minutes, chlorodiphenylphosphine (77 µL, 62.5 µmol, 1.00 equiv.) was added dropwise. The reaction was slowly allowed to warm up to r.t. over 2 h. In an argon counter-current (tht)gold(I)chloride (112 mg, 348 µmol, 1.00 equiv.) was added to the mixture and stirred for 2 hours at r.t. in the dark. After removal of the solvent the crude residue was purified by column chromatography to afford the product as a white solid (115 mg, 53%).

R_f = 0.18 (pentane/ethyl acetate 5:1).

^{1}H NMR (500 MHz, Chloroform-*d*) δ 7.61 – 7.52 (m, 4H), 7.51 – 7.44 (m, 5H), 7.37 (td, *J* = 7.7, 2.8 Hz, 2H), 6.65 – 6.57 (m, 3H), 6.52 (dd, *J* = 7.7, 5.4 Hz, 1H), 6.22 (dd, *J* = 8.0, 1.6 Hz, 1H), 5.93 (dd, *J* = 14.8, 1.8 Hz, 1H), 3.82 (ddd, *J* = 12.8, 10.7, 4.0 Hz, 1H), 3.73 (dddd, *J* = 12.1, 10.8, 2.8, 1.3 Hz, 1H), 3.13 – 2.84 (m, 5H), 2.71 (ddd, *J* = 13.3, 10.3, 5.0 Hz, 1H).

^{13}C NMR (101 MHz, $CDCl_3$) δ 144.5, 144.4, 140.6, 140.5, 139.3, 137.0, 136.9, 136.4, 136.4, 136.3, 136.2, 135.2, 135.1, 134.1, 134.0, 133.0, 132.4, 132.2, 132.1, 132.0, 132.0, 131.9, 131.9, 131.9, 130.6, 130.1, 129.3, 129.2, 129.0, 128.9, 128.7, 126.4, 125.9, 35.1, 35.1, 34.7, 34.6.

IR (ATR, ṽ) = 2948, 2922, 2890, 2850, 1479, 1435, 1409, 1392, 1181, 1099, 1027, 997, 899, 847, 793, 747, 718, 708, 691, 637, 579, 547, 507, 490, 463, 442, 429 cm^{-1}.

4-Diphenylphosphine gold(I) chloride[2.2]paracyclophane (192)

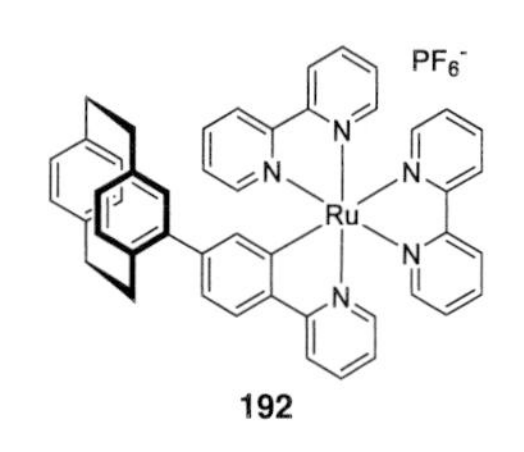

A solution of [Ru(bpy)$_2$Cl$_2$] (100 mg, 199 μmol, 1.00 equiv.), 4-phenylpyridine[2.2]paracyclophane (216 mg, 597 μmol, 3.00 equiv.) and silver tetrafluoroborate (80.2 mg, 412 μmol, 2.07 equiv.) in dichloromethane (16 mL) was heated to reflux for 5 h. The solution was cooled to room temperature, filtered through Celite and the solvent was removed under reduced pressure. The solid was taken up in acetonitrile and ammonium hexafluorophosphate (97.3 mg, 597 μmol, 3.00 equiv.) in methanol (1.40 mL) was added. The solvent was removed under reduced pressure and the crude solid was purified by column chromatography (silica, DCM/MeCN, 15:1) to yield [Ru(bpy)$_2$(Pc-ppy)]PF$_6$ (85.0 mg, 46%) as a diastereomeric mixture of a deep red solid.

^{1}H NMR [diastereomeric mixture – double signal set] (500 MHz, Acetonitrile-*d*$_3$) δ 8.52 – 8.48 (m, 2H), 8.47 (dd, *J* = 8.3, 1.3 Hz, 1H), 8.45 – 8.39 (m, 3H), 8.27 – 8.21 (m, 3H), 8.18 – 8.15 (m, 1H), 8.06 – 7.89 (m, 9H), 7.88 – 7.75 (m, 7H), 7.75 – 7.65 (m, 4H), 7.57 (t, *J* = 5.5 Hz, 2H), 7.47 (qd, *J* = 5.9, 5.3, 2.7 Hz, 2H), 7.35 – 7.23 (m, 5H), 7.19 (ddd, *J* = 7.2, 5.5, 1.3 Hz, 1H), 6.88 (q, *J* = 7.9, 7.5 Hz, 2H), 6.68 (d, *J* = 23.6 Hz, 2H), 6.57 (ddd, *J* = 7.8, 4.5, 1.9 Hz, 2H), 6.53 – 6.41 (m, 6H), 6.37 (d, *J* = 1.8 Hz, 1H), 6.33 (dd, *J* = 7.9, 1.9 Hz, 1H), 6.16 (d, *J* = 1.8 Hz, 1H), 6.06 (dd, *J* = 7.8, 1.9 Hz, 1H), 6.00 (dd, *J* = 7.8, 1.9 Hz, 1H), 5.94 (dd, *J* = 7.8, 1.9 Hz, 1H), 3.36 – 3.20 (m, 2H), 3.10 – 2.93 (m, 6H), 2.89 – 2.68 (m, 6H), 2.46 – 2.37 (m, 1H), 2.30 – 2.21 (m, 1H).

^{13}C NMR [diastereomeric mixture – double signal set] (126 MHz, CD$_3$CN) δ 159.0, 159.0, 158.2, 158.1, 158.0, 157.9, 156.3, 156.2, 155.5, 155.4, 151.7, 151.6, 151.4, 151.3, 151.1, 151.0, 143.7, 143.6, 140.6, 140.6, 140.6, 140.5, 140.4, 138.0, 137.8, 137.4, 136.8, 136.8, 136.5, 136.5, 136.0, 136.0, 134.9, 134.9, 134.7, 133.9, 133.8, 133.6, 133.5, 133.0, 132.9, 132.7, 132.7, 132.5, 132.3, 130.6, 130.4, 129.9, 129.2, 128.2, 128.1, 127.5, 127.4, 127.2, 126.8, 126.7, 124.5, 124.4, 124.2, 124.1, 124.1, 124.0, 123.8, 123.6, 123.4, 122.2, 120.1, 120.1, 35.9, 35.9, 35.7, 35.6, 35.4, 35.3, 35.3, 35.2.

IR (ATR, cm^{-1}) $\tilde{\nu}$ = 2921, 2851, 1596, 1572, 1554, 1459, 1441, 1419, 1360, 1309, 1256, 1242, 1157, 1057, 1027, 1014, 1001, 878, 832, 782, 755, 727, 673, 652, 555, 518, 510, 487, 419.

HRMS ($C_{47}H_{38}N_5Ru$) calc.: 774.2165; found: 774.2153.

4-Diphenylphosphoryl-12-(4-(2'-pyridyl)phenyl)[2.2]paracyclophane (179)

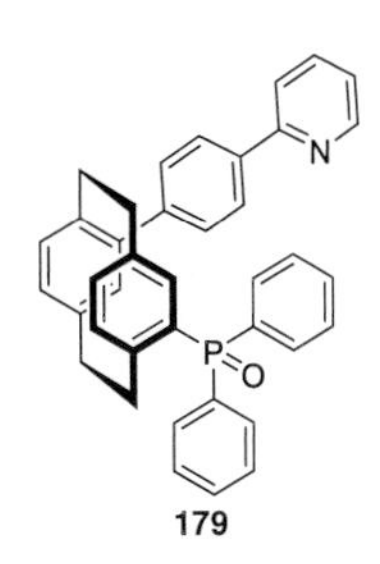

In a vial with a magnetic stirring bar, 4-bromo-12-diphenylphosphoryl[2.2]paracyclophane (600 mg, 1.20 mmol, 1.00 equiv.), (4-pyridine-2-ylphenyl)boronic acid (368 mg, 1.80 mmol, 1.50 equiv.), palladium(II) acetate (2.76 mg, 12.0 µmol, 1 mol%), rac-AntPhos (9.12 mg, 25.0 µmol, 2 mol%) and potassium phosphate (784 mg, 3.70 mmol, 3.00 equiv.) were placed. The vial was capped and evacuated and backfilled with argon three times. Then, degassed dioxane (2.4 mL) was added with a syringe. The mixture was stirred and heated to 110 °C overnight. The next day, the mixture was passed over a Celite pad and the solvent removed under reduced pressure. The crude mixture was subjected to column chromatography (silica, pentane/ethyl acetate 2:1 to 1:5) to yield the title compound as an off-white solid (410 mg, 59%).

^{1}**H NMR** (300 MHz, Chloroform-*d*) δ 8.77 – 8.68 (m, 1H), 7.91 (t, *J* = 9.0 Hz, 4H), 7.82 – 7.62 (m, 6H), 7.62 – 7.33 (m, 6H), 7.22 (dd, *J* = 4.0, 1.6 Hz, 2H), 6.78 – 6.68 (m, 3H), 6.68 – 6.59 (m, 2H), 3.70 (dd, *J* = 13.4, 9.1 Hz, 1H), 3.42 (dd, *J* = 11.8, 6.0 Hz, 1H), 3.31 (t, *J* = 11.6 Hz, 1H), 3.07 (t, *J* = 11.1 Hz, 1H), 2.97 – 2.68 (m, 3H), 2.13 (dt, *J* = 12.8, 8.8 Hz, 1H).

4-(Diphenylphosphinyl)-12-(4-(2'-pyridyl)phenyl)[2.2]paracyclophane (182)

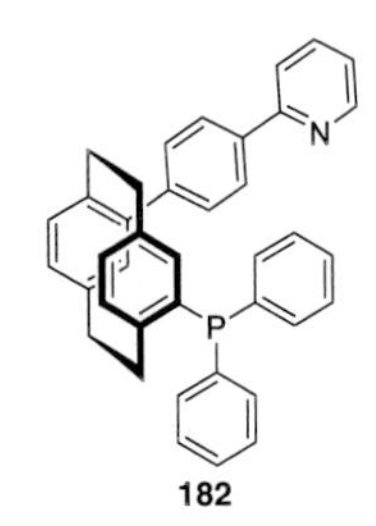

According to **GP phosphine oxide reduction** 4-Diphenylphosphoryl-12-(4-(2'-pyridyl)phenyl)[2.2]paracyclophane (**163**) was reduced to the corresponding phosphine. The crude was used without further purification.

4-(Diphenylphosphinyl gold(I) chloride)-12-(4-(2'-pyridyl)phenyl)[2.2]para-cyclophane (181)

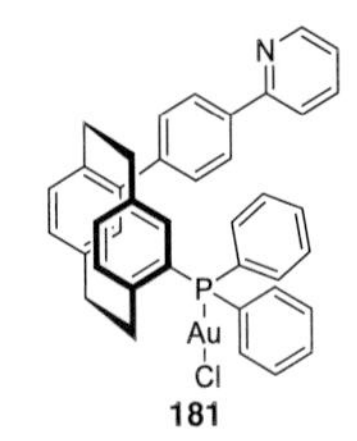

According to **GP phosphine auration**, the title product was obtained after column chromatography as a white solid (360 mg, 63%).

^{1}H NMR (500 MHz, Chloroform-*d*) δ 8.73 (tt, *J* = 3.7, 1.8 Hz, 1H), 8.02 – 7.95 (m, 2H), 7.88 – 7.72 (m, 4H), 7.61 (ddt, *J* = 13.9, 9.0, 1.7 Hz, 2H), 7.58 – 7.47 (m, 6H), 7.47 – 7.42 (m, 1H), 7.38 (qd, *J* = 7.3, 4.0 Hz, 2H), 7.30 – 7.24 (m, 1H), 6.83 – 6.75 (m, 1H), 6.75 – 6.67 (m, 3H), 6.19 (dd, *J* = 14.5, 1.5 Hz, 1H), 3.87 (ddt, *J* = 16.1, 8.6, 4.3 Hz, 1H), 3.65 (dd, *J* = 13.7, 9.2 Hz, 1H), 3.53 – 3.44 (m, 1H), 3.13 (ddd, *J* = 13.2, 11.1, 1.7 Hz, 1H), 3.00 – 2.84 (m, 3H), 2.82 – 2.72 (m, 1H).

^{13}C NMR (126 MHz, $CDCl_3$) δ 157.3, 157.2, 149.9, 149.8, 144.4, 144.3, 141.7, 141.7, 141.1, 140.8, 140.6, 140.5, 139.7, 139.6, 138.1, 137.4, 137.0, 137.0, 137.0, 137.0, 136.3, 136.2, 136.1, 136.0, 135.6, 135.5, 135.4, 135.3, 135.0, 134.2, 134.1, 133.8, 133.7, 133.7, 133.6, 132.7, 132.7, 132.3, 132.2, 132.2, 132.0, 132.0, 131.9, 131.9, 131.8, 131.4, 130.8, 130.2, 129.9, 129.4, 129.4, 129.3, 129.2, 129.2, 129.1, 129.1, 129.1, 128.6, 127.3, 127.2, 125.5, 125.0, 122.3, 122.2, 120.7, 120.7, 60.5, 35.3, 35.0, 34.9, 34.9, 34.4, 33.4, 33.1, 30.6, 21.2, 14.3.

HRMS ($C_{39}H_{32}AuClNP + H^{+}$) calc. 778.1699; found 778.1695.

4-(((Bis(2,2'-bipyridyl)4-(2'-pyridyl)phenyl) ruthenium (II))-16-(chloro-(diphenylphosphinyl)gold(I))[2.2]paracyclophane (183)

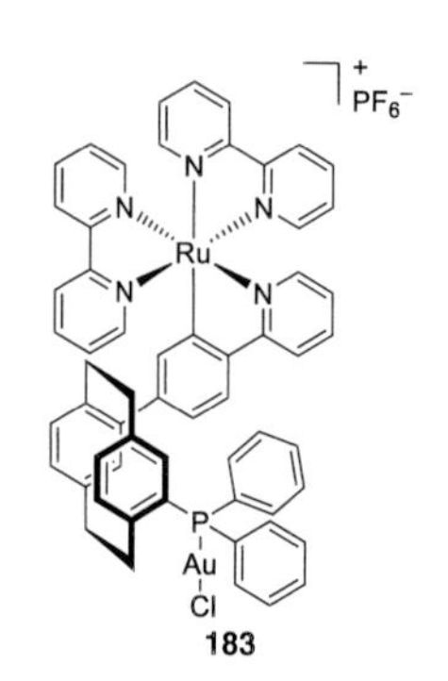

In a vial under argon 4-(4-(2'-pyridyl)phenyl))12-(chloro(diphenylphosphinyl)-gold(I))[2.2]paracyclophane (305 mg, 392 μmol, 1.00 equiv.), (Benzene)ruthenium dichloride dimer (98.4 mg, 196 μmol, 0.500 equiv.), sodium hydroxide (15.7 mg, 392 μmol, 1.00 equiv.) and potassium hexafluorophosphate (144 mg, 784 μmol, 2.00 equiv.) were dissolved in degassed acetonitrile (10.0 mL) and stirred at room temperature overnight. The next day, 2,2'-bipyridine (122 mg, 784 μmol, 2.00 equiv.) was added and the mixture was refluxed overnight. The solution changed from green to dark red. The solvent was removed, and the crude purified on silica gel with dichloromethane/acetonitrile 10:1. The first red band was collected to afford the title compound as a dark red solid (30.0 mg, 6%).

^{1}H NMR [diastereomeric mixture – double signal set] (500 MHz, CD_3CN) δ 8.70 (dtt, J = 4.7, 2.6, 1.3 Hz, 1H), 8.45 (dt, J = 8.3, 1.1 Hz, 2H), 8.40 – 8.27 (m, 5H), 8.23 – 8.01 (m, 5H), 8.01 – 7.71 (m, 16H), 7.71 – 7.36 (m, 26H), 7.38 – 7.28 (m, 2H), 7.27 – 7.21 (m, 1H), 7.18 (dddd, J = 10.6, 5.8, 3.6, 1.2 Hz, 3H), 7.11 (dddd, J = 14.6, 7.4, 5.7, 1.4 Hz, 2H), 7.00 – 6.93 (m, 1H), 6.88 (dd, J = 11.5, 7.7 Hz, 1H), 6.78 (dddt, J = 10.3, 8.5, 4.7, 2.3 Hz, 4H), 6.73 – 6.66 (m, 2H), 6.65 – 6.59 (m, 3H), 6.47 – 6.26 (m, 1H), 6.25 – 6.17 (m, 1H), 6.17 – 6.07 (m, 1H), 6.06 – 5.84 (m, 1H), 3.83 – 3.54 (m, 4H), 3.47 – 3.20 (m, 3H), 3.18 – 3.02 (m, 3H), 3.01 – 2.76 (m, 4H), 2.64 – 2.54 (m, 1H), 2.46 (ddd, J = 13.4, 9.5, 8.3 Hz, 1H).

^{13}C NMR [diastereomeric mixture – double signal set] (126 MHz, CD_3CN) δ 167.8, 158.5, 158.4, 157.9, 157.7, 157.2, 157.2, 156.1, 155.4, 155.2, 154.8, 151.3, 151.2, 151.1, 151.0, 150.7, 150.4, 150.2, 145.7, 145.6, 145.1, 145.0, 145.0, 144.9, 142.8, 142.1, 142.0, 141.4, 141.3, 141.1, 141.0, 140.8, 140.8, 140.7, 140.6, 140.5, 140.4, 139.0, 138.8, 138.8, 138.6, 138.4, 138.2, 138.2, 138.1, 138.1, 138.0, 138.0, 137.8, 137.6, 137.3, 137.2, 137.2, 137.1, 137.0, 136.9, 136.8, 136.8, 136.7, 136.7, 136.6, 136.5, 136.2, 136.1, 135.9, 135.8, 135.8, 134.6, 134.6, 134.5, 134.5, 134.4, 134.4, 134.4, 134.3, 134.2, 134.1, 134.1, 133.7, 133.7, 133.6, 133.4, 133.3, 133.2, 133.2, 133.1, 133.1, 132.6, 132.5, 132.4, 132.0, 131.7, 131.2, 131.1, 130.8, 130.7, 130.7, 130.6, 130.5, 130.5,

130.4, 130.4, 130.3, 130.3, 130.2, 130.2, 130.1, 130.1, 130.1, 129.6, 128.1, 128.0, 128.0, 127.9, 127.9, 127.8, 127.8, 127.3, 127.2, 127.0, 126.3, 126.1, 125.6, 124.8, 124.8, 124.5, 124.4, 124.2, 124.0, 123.9, 123.7, 123.5, 123.5, 123.3, 123.3, 122.1, 122.0, 121.3, 121.3, 120.1, 120.0, 79.1, 36.1, 35.9, 35.7, 35.6, 35.4, 35.3, 35.3, 35.2, 35.1, 34.6, 34.2, 34.0, 33.8, 33.7, 33.7, 33.6.

^{31}P NMR (162 MHz, $CDCl_3$) δ 32.0.

HRMS ($C_{59}H_{47}AuClN_5PRu$) calc. 1190.1961; found 1190.1966.

6 REFERENCES

[1] Z. Hassan, E. Spuling, D. M. Knoll, J. Lahann, S. Bräse, *Chem. Soc. Rev.* **2018**, *47*, 6947–6963.

[2] D. J. Cram, J. M. Cram, *Acc. Chem. Res.* **1971**, *4*, 204–213.

[3] I. Majerz, T. Dziembowska, *J. Phys. Chem. A* **2016**, *120*, 8138–8147.

[4] D. M. Knoll, H. Šimek, Z. Hassan, S. Bräse, *Eur. J. Org. Chem.* **2019**, *36*, 6198–6202.

[5] D. S. Seferos, S. A. Trammell, G. C. Bazan, J. G. Kushmerick, *Proc. Natl. Acad. Sci. U. S. A.* **2005**, *102*, 8821–8825.

[6] M. Wielopolski, A. Molina-Ontoria, C. Schubert, J. T. Margraf, E. Krokos, J. Kirschner, A. Gouloumis, T. Clark, D. M. Guldi, N. Martín, *J. Am. Chem. Soc.* **2013**, *135*, 10372–10381.

[7] A. Molina-Ontoria, M. Wielopolski, J. Gebhardt, A. Gouloumis, T. Clark, D. M. Guldi, N. Martín, *J. Am. Chem. Soc.* **2011**, *133*, 2370–2373.

[8] C. J. Brown, A. C. Farthing, *Nature* **1949**, *164*, 915–916.

[9] F. Vögtle, P. Neumann, *Tetrahedron Lett.* **1969**, *10*, 5329–5334.

[10] H. Hopf, *Isr. J. Chem.* **2012**, *52*, 18–19.

[11] Y. L. Yeh, W. F. Gorham, *J. Org. Chem.* **1969**, *34*, 2366–2370.

[12] W. F. Gorham, *J. Polym. Sci. [A1]* **1966**, *4*, 3027–3039.

[13] J. P. Seymour, Y. M. Elkasabi, H.-Y. Chen, J. Lahann, D. R. Kipke, *Biomaterials* **2009**, *30*, 6158–6167.

[14] C. P. Tan, H. G. Craighead, *Materials* **2010**, *3*, DOI 10.3390/ma3031803.

[15] J. Lahann, M. Balcells, H. Lu, T. Rodon, K. F. Jensen, R. Langer, in *Micro Total Anal. Syst. 2002* (Eds.: Y. Baba, S. Shoji, A. van den Berg), Springer Netherlands, **2002**, pp. 443–445.

[16] G. Maggioni, A. Campagnaro, S. Carturan, A. Quaranta, *Sel. Publ. 22nd Space Photovolt. Res. Technol. SPRAT Conf.* **2013**, *108*, 27–37.

[17] S. Kuppusami, R. H. Oskouei, *Univers. J. Biomed. Eng.* **2015**, *3*, 9–14.

[18] J. Paradies, *Synthesis* **2011**, *2011*, 3749–3766.

[19] S. Bräse, S. Dahmen, S. Höfener, F. Lauterwasser, M. Kreis, R. E. Ziegert, *Synlett* **2004**, *2004*, 2647–2669.

[20] M. Busch, M. Cayir, M. Nieger, W. R. Thiel, S. Bräse, *Eur. J. Org. Chem.* **2013**, *2013*, 6108–6123.

[21] S. E. Gibson, J. D. Knight, *Org. Biomol. Chem.* **2003**, *1*, 1256–1269.

[22] B. Jiang, Y. Lei, X.-L. Zhao, *J. Org. Chem.* **2008**, *73*, 7833–7836.

[23] P. W. Dyer, P. J. Dyson, S. L. James, C. M. Martin, P. Suman, *Organometallics* **1998**, *17*, 4344–4346.

[24] P. J. Pye, K. Rossen, R. A. Reamer, R. P. Volante, P. J. Reider, *Tetrahedron Lett.* **1998**, *39*, 4441–4444.

[25] C. Braun, S. Bräse, L. L. Schafer, *Eur. J. Org. Chem.* **2017**, *2017*, 1760–1764.

[26] S. Kitagaki, S. Murata, K. Asaoka, K. Sugisaka, C. Mukai, N. Takenaga, K. Yoshida, *Chem. Pharm. Bull. (Tokyo)* **2018**, *66*, 1006–1014.

[27] X.-L. Hou, X.-W. Wu, L.-X. Dai, B.-X. Cao, J. Sun, *Chem. Commun.* **2000**, 1195–1196.

[28] P. An, Y. Huo, Z. Chen, C. Song, Y. Ma, *Org. Biomol. Chem.* **2017**, *15*, 3202–3206.

[29] M. Gon, Y. Morisaki, Y. Chujo, *J. Mater. Chem. C* **2015**, *3*, 521–529.

[30] G. Zhou, W.-Y. Wong, X. Yang, *Chem. – Asian J.* **2011**, *6*, 1706–1727.

[31] M. Jamshidi, R. Yousefi, S. M. Nabavizadeh, M. Rashidi, M. G. Haghighi, A. Niazi, A.-A. Moosavi-Movahedi, *Int. J. Biol. Macromol.* **2014**, *66*, 86–96.

[32] I. M. El-Mehasseb, M. Kodaka, T. Okada, T. Tomohiro, K. Okamoto, H. Okuno, *J. Inorg. Biochem.* **2001**, *84*, 157–158.

[33] S. Liu, H. Sun, Y. Ma, S. Ye, X. Liu, X. Zhou, X. Mou, L. Wang, Q. Zhao, W. Huang, *J. Mater. Chem.* **2012**, *22*, 22167–22173.

[34] H. Huang, P. Zhang, H. Chen, L. Ji, H. Chao, *Chem. – Eur. J.* **2015**, *21*, 715–725.

[35] H. Hopf, A. A. Aly, V. N. Swaminathan, L. Ernst, I. Dix, P. G. Jones, *Eur. J. Org. Chem.* **2005**, *2005*, 68–71.

[36] J. R. Fulton, J. E. Glover, L. Kamara, G. J. Rowlands, *Chem. Commun.* **2011**, *47*, 433–435.

[37] L.-C. Campeau, S. Rousseaux, K. Fagnou, *J. Am. Chem. Soc.* **2005**, *127*, 18020–18021.

[38] Q. Chai, C. Song, Z. Sun, Y. Ma, C. Ma, Y. Dai, M. B. Andrus, *Tetrahedron Lett.* **2006**, *47*, 8611–8615.

[39] U. Wörsdörfer, F. Vögtle, M. Nieger, M. Waletzke, S. Grimme, F. Glorius, A. Pfaltz, *Synthesis* **1999**, *1999*, 597–602.

[40] J. M. R. Narayanam, J. W. Tucker, C. R. J. Stephenson, *J. Am. Chem. Soc.* **2009**, *131*, 8756–8757.

[41] D. A. Nicewicz, D. W. C. MacMillan, *Science* **2008**, *322*, 77.

[42] M. A. Ischay, M. E. Anzovino, J. Du, T. P. Yoon, *J. Am. Chem. Soc.* **2008**, *130*, 12886–12887.

[43] Y. Chen, A. S. Kamlet, J. B. Steinman, D. R. Liu, *Nat. Chem.* **2011**, *3*, 146.

[44] H. Cano-Yelo, A. Deronzier, *J. Photochem.* **1987**, *37*, 315–321.

[45] L. Marzo, S. K. Pagire, O. Reiser, B. König, *Angew. Chem. Int. Ed.* **2018**, *57*, 10034–10072.

[46] C. R. Stephenson, T. P. Yoon, D. W. MacMillan, *Visible Light Photocatalysis in Organic Chemistry*, John Wiley & Sons, **2018**.

[47] H. H. Jaffe, A. L. Miller, *J. Chem. Educ.* **1966**, *43*, 469.

[48] M. Reckenthäler, A. G. Griesbeck, *Adv. Synth. Catal.* **2013**, *355*, 2727–2744.

[49] Y. Pellegrin, F. Odobel, *Artif. Photosynth. Photosynth. Artif.* **2017**, *20*, 283–295.

[50] K. L. Skubi, T. R. Blum, T. P. Yoon, *Chem. Rev.* **2016**, *116*, 10035–10074.

[51] A. S. K. Hashmi, G. J. Hutchings, *Angew. Chem. Int. Ed.* **2006**, *45*, 7896–7936.

[52] M. N. Hopkinson, A. Tlahuext-Aca, F. Glorius, *Acc. Chem. Res.* **2016**, *49*, 2261–2272.

[53] H. Li, C. Shan, C.-H. Tung, Z. Xu, *Chem. Sci.* **2017**, *8*, 2610–2615.

[54] H. Li, Z. Cheng, C.-H. Tung, Z. Xu, *ACS Catal.* **2018**, *8*, 8237–8243.

[55] D. S. Hamilton, D. A. Nicewicz, *J. Am. Chem. Soc.* **2012**, *134*, 18577–18580.

[56] P. R. Ogilby, C. S. Foote, *J. Am. Chem. Soc.* **1983**, *105*, 3423–3430.

[57] W. Adam, A. Griesbeck, E. Staab, *Tetrahedron Lett.* **1986**, *27*, 2839–2842.

[58] Muller P., *Pure Appl. Chem.* **1994**, *66*, 1077.

[59] J. Zyss, I. Ledoux, S. Volkov, V. Chernyak, S. Mukamel, G. P. Bartholomew, G. C. Bazan, *J. Am. Chem. Soc.* **2000**, *122*, 11956–11962.

[60] Y. Morisaki, Y. Chujo, *Macromolecules* **2003**, *36*, 9319–9324.

[61] E. Spuling, N. Sharma, I. D. W. Samuel, E. Zysman-Colman, S. Bräse, *Chem. Commun.* **2018**, *54*, 9278–9281.

[62] M. R. D. Gatus, M. Bhadbhade, B. A. Messerle, *Dalton Trans.* **2017**, *46*, 14406–14419.

[63] R. H. Crabtree, *J. Organomet. Chem.* **2005**, *690*, 5451–5457.

[64] D. M. Knoll, Synthetic Approaches to Disparately Pseudo-Para Substituted [2.2]Paracyclophanes, Master's Thesis, KIT, **2016**.

[65] J. F. Hartwig, *Organotransition Metal Chemistry: From Bonding to Catalysis*, Univ Science Books, **2010**.

[66] D. G. Gilheany, *Chem. Rev.* **1994**, *94*, 1339–1374.

[67] S. A. Shahzad, M. A. Sajid, Z. A. Khan, D. Canseco-Gonzalez, *Synth. Commun.* **2017**, *47*, 735–755.

[68] W. Zi, F. D. Toste, *Chem. Soc. Rev.* **2016**, *45*, 4567–4589.

[69] A. Tlahuext-Aca, M. N. Hopkinson, R. A. Garza-Sanchez, F. Glorius, *Chem. – Eur. J.* **2016**, *22*, 5909–5913.

[70] J. Um, H. Yun, S. Shin, *Org. Lett.* **2016**, *18*, 484–487.

[71] X. Ren, Z. Lu, *Chin. J. Catal.* **2019**, *40*, 1003–1019.

[72] B. Alcaide, P. Almendros, E. Busto, A. Luna, *Adv. Synth. Catal.* **2016**, *358*, 1526–1533.

[73] M. J. McKeage, L. Maharaj, S. J. Berners-Price, *Coord. Chem. Rev.* **2002**, *232*, 127–135.

[74] C. Bolm, K. Wenz, G. Raabe, *J. Organomet. Chem.* **2002**, *662*, 23–33.

[75] E. Ferrer Flegeau, C. Bruneau, P. H. Dixneuf, A. Jutand, *J. Am. Chem. Soc.* **2011**, *133*, 10161–10170.

[76] H. A. Younus, N. Ahmad, W. Su, F. Verpoort, *Coord. Chem. Rev.* **2014**, *276*, 112–152.

[77] G. C. Fortman, S. P. Nolan, *Chem. Soc. Rev.* **2011**, *40*, 5151–5169.

[78] A. C. Hillier, G. A. Grasa, M. S. Viciu, H. M. Lee, C. Yang, S. P. Nolan, *J. Organomet. Chem.* **2002**, *653*, 69–82.

[79] N. T. S. Phan, M. Van Der Sluys, C. W. Jones, *Adv. Synth. Catal.* **2006**, *348*, 609–679.

[80] B. Basu, K. Biswas, S. Kundu, S. Ghosh, *Green Chem.* **2010**, *12*, 1734–1738.

[81] N. V. Atman, R. P. Zhuravsky, D. Y. Antonov, E. V. Sergeeva, I. A. Godovikov, Z. A. Starikova, A. V. Vologzhanina, *Eur. J. Org. Chem.* **2015**, *2015*, 325–330.

[82] Norio. Miyaura, Akira. Suzuki, *Chem. Rev.* **1995**, *95*, 2457–2483.

[83] T. E. Barder, S. D. Walker, J. R. Martinelli, S. L. Buchwald, *J. Am. Chem. Soc.* **2005**, *127*, 4685–4696.

[84] A. J. J. Lennox, G. C. Lloyd-Jones, *Chem. Soc. Rev.* **2013**, *43*, 412–443.

[85] J. Huang, S. P. Nolan, *J. Am. Chem. Soc.* **1999**, *121*, 9889–9890.

[86] E. Negishi, *J. Organomet. Chem.* **2002**, *653*, 34–40.

[87] P. KUS, A. ZEMANEK, *Chem. Informationsdienst* **1986**, *17*, no-no.

[88] C. Braun, E. Spuling, N. B. Heine, M. Cakici, M. Nieger, S. Bräse, *Adv. Synth. Catal.* **2016**, *358*, 1664–1670.

[89] D. Haas, J. M. Hammann, R. Greiner, P. Knochel, *ACS Catal.* **2016**, *6*, 1540–1552.

[90] S. Thapa, A. S. Vangala, R. Giri, *Synthesis* **2016**, *48*, 504–511.

[91] J.-M. Bégouin, C. Gosmini, *J. Org. Chem.* **2009**, *74*, 3221–3224.

[92] J.-M. Begouin, M. Rivard, C. Gosmini, *Chem. Commun.* **2010**, *46*, 5972–5974.

[93] S. Huo, *Org. Lett.* **2003**, *5*, 423–425.

[94] L. Wang, Z.-X. Wang, *Org. Lett.* **2007**, *9*, 4335–4338.

[95] S. Son, G. C. Fu, *J. Am. Chem. Soc.* **2008**, *130*, 2756–2757.

[96] A. Joshi-Pangu, M. Ganesh, M. R. Biscoe, *Org. Lett.* **2011**, *13*, 1218–1221.

[97] C.-Y. Huang, A. G. Doyle, *J. Am. Chem. Soc.* **2012**, *134*, 9541–9544.

[98] J. K. Stille, *Angew. Chem. Int. Ed. Engl.* **1986**, *25*, 508–524.

[99] C. Cordovilla, C. Bartolomé, J. M. Martínez-Ilarduya, P. Espinet, *ACS Catal.* **2015**, *5*, 3040–3053.

[100] Farina Vittorio, *Pure Appl. Chem.* **1996**, *68*, 73.

[101] V. Farina, B. Krishnan, D. R. Marshall, G. P. Roth, *J. Org. Chem.* **1993**, *58*, 5434–5444.

[102] C. Braun, Synthese Und Anwendung Planar-Chiraler [2.2]Paracyclophanliganden Mit N-Donorfunktionen Und Deren Übergangsmetallkomplexe Zur Synthese Substituierter Amine, PhD Thesis, KIT, **2017**.

[103] C. Braun, M. Nieger, S. Bräse, *Chem.- Eur. J.* **2017**, *23*, 16452–16455.

[104] F.-S. Han, *Chem. Soc. Rev.* **2013**, *42*, 5270–5298.

[105] S. E. Hooshmand, B. Heidari, R. Sedghi, R. S. Varma, *Green Chem.* **2019**, *21*, 381–405.

[106] A. J. Roche, B. Canturk, *Org. Biomol. Chem.* **2005**, *3*, 515–519.

[107] G. A. Molander, N. Ellis, *Acc. Chem. Res.* **2007**, *40*, 275–286.

[108] E. Vedejs, R. W. Chapman, S. C. Fields, S. Lin, M. R. Schrimpf, *J. Org. Chem.* **1995**, *60*, 3020–3027.

[109] S. Darses, J.-P. Genêt, J.-L. Brayer, J.-P. Demoute, *Tetrahedron Lett.* **1997**, *38*, 4393–4396.

[110] G. A. Molander, B. Biolatto, *J. Org. Chem.* **2003**, *68*, 4302–4314.

[111] C. F. R. A. C. Lima, A. S. M. C. Rodrigues, V. L. M. Silva, A. M. S. Silva, L. M. N. B. F. Santos, *ChemCatChem* **2014**, *6*, 1291–1302.

[112] A. J. J. Lennox, G. C. Lloyd-Jones, *J. Am. Chem. Soc.* **2012**, *134*, 7431–7441.

[113] E. Vedejs, R. W. Chapman, S. C. Fields, S. Lin, M. R. Schrimpf, *J. Org. Chem.* **1995**, *60*, 3020–3027.

[114] V. Bagutski, A. Ros, V. K. Aggarwal, *Tetrahedron* **2009**, *65*, 9956–9960.

[115] M. Butters, J. N. Harvey, J. Jover, A. J. J. Lennox, G. C. Lloyd-Jones, P. M. Murray, *Angew. Chem. Int. Ed.* **2010**, *49*, 5156–5160.

[116] S. Darses, G. Michaud, J.-P. Genêt, *Eur. J. Org. Chem.* **1999**, *1999*, 1875–1883.

[117] A. J. J. Lennox, G. C. Lloyd-Jones, *Angew. Chem. Int. Ed.* **2012**, *51*, 9385–9388.

[118] W. Ren, J. Li, D. Zou, Y. Wu, Y. Wu, *Tetrahedron* **2012**, *68*, 1351–1358.

[119] E. Vedejs, R. W. Chapman, S. C. Fields, S. Lin, M. R. Schrimpf, *J. Org. Chem.* **1995**, *60*, 3020–3027.

[120] G. A. Molander, B. Canturk, *Angew. Chem. Int. Ed.* **2009**, *48*, 9240–9261.

[121] C. Liu, X. Li, Z. Gao, X. Wang, Z. Jin, *Tetrahedron* **October 6**, *71*, 3954–3959.

[122] R. A. Batey, T. D. Quach, *Tetrahedron Lett.* **12**, *42*, 9099–9103.

[123] T. D. Quach, R. A. Batey, *Org. Lett.* **2003**, *5*, 1381–1384.

[124] G. A. Molander, D. L. Sandrock, *Curr. Opin. Drug Discov. Devel.* **2009**, *12*, 811–823.

[125] Y. A. Cho, D.-S. Kim, H. R. Ahn, B. Canturk, G. A. Molander, J. Ham, *Org. Lett.* **2009**, *11*, 4330–4333.

[126] W. Ren, J. Li, D. Zou, Y. Wu, Y. Wu, *Tetrahedron* **2012**, *68*, 1351–1358.

[127] G. A. Molander, *J. Org. Chem.* **2015**, *80*, 7837–7848.

[128] S. Darses, G. Michaud, J. Genêt, *Eur. J. Org. Chem.* **1999**, *1999*, 1875–1883.

[129] B. Buck, A. Mascioni, L. Que, G. Veglia, *J. Am. Chem. Soc.* **2003**, *125*, 13316–13317.

[130] G. A. Molander, D. L. Sandrock, *Curr. Opin. Drug Discov. Devel.* **2009**, *12*, 811–823.

[131] D. M. Knoll, S. Bräse, *ACS Omega* **2018**, *3*, 12158–12162.

[132] J. Schlindwein, Bachelor's thesis - unpublished work, KIT, **2018**.

[133] Moss G. P., Smith P. A. S., Tavernier D., *Pure Appl. Chem.* **1995**, *67*, 1307.

[134] "Chlorophyll," can be found under http://www.chm.bris.ac.uk/motm/chlorophyll/chlorophyll_h.htm, **2019**.

[135] G. M. Mason, L. G. Trudell, J. F. Branthaver, *Org. Geochem.* **1989**, *14*, 585–594.

[136] "porphyrin | Search Online Etymology Dictionary," can be found under https://www.etymonline.com/search?q=porphyrin, **2019**.

[137] J. Kou, D. Dou, L. Yang, *Oncotarget* **2017**, *8*, 81591–81603.

[138] C. M. Davis, K. Ohkubo, A. D. Lammer, D. S. Kim, Y. Kawashima, J. L. Sessler, S. Fukuzumi, *Chem. Commun.* **2015**, *51*, 9789–9792.

[139] K. A. Nielsen, W.-S. Cho, G. H. Sarova, B. M. Petersen, A. D. Bond, J. Becher, F. Jensen, D. M. Guldi, J. L. Sessler, J. O. Jeppesen, *Angew. Chem. Int. Ed.* **2006**, *45*, 6848–6853.

[140] X. Huang, J. T. Groves, *Chem. Rev.* **2018**, *118*, 2491–2553.

[141] G. Wilkinson, R. D. Gillard, *Comprehensive Coordination Chemistry The Synthesis, Reactions, Properties and Applications of Coordination Compounds V3 Main Group and Early Transition Elements*, Pergamon Press, United Kingdom, **1987**.

[142] P. Rothemund, *J. Am. Chem. Soc.* **1935**, *57*, 2010–2011.

[143] A. D. Adler, F. R. Longo, J. D. Finarelli, J. Goldmacher, J. Assour, L. Korsakoff, *J. Org. Chem.* **1967**, *32*, 476–476.

[144] J. S. Lindsey, H. C. Hsu, I. C. Schreiman, *Tetrahedron Lett.* **1986**, *27*, 4969–4970.

[145] P. D. Rao, S. Dhanalekshmi, B. J. Littler, J. S. Lindsey, *J. Org. Chem.* **2000**, *65*, 7323–7344.

[146] J. S. Lindsey, *Acc. Chem. Res.* **2009**, *43*, 300–311.

[147] T. B. Wiesner, Synthesis and Characterisation of Porphyrin-[2.2]Paracyclophane-Conjugates, Master's Thesis, KIT, **2017**.

[148] C. Schissler, PhD Thesis, Unpublished Results, KIT, **tbd**.

[149] C. Braun, M. Nieger, W. R. Thiel, S. Bräse, *Chem. - Eur. J.* **2017**, *23*, 15474–15483.

[150] D. K. Whelligan, C. Bolm, *J. Org. Chem.* **2006**, *71*, 4609–4618.

[151] A. Marchand, A. Maxwell, B. Mootoo, A. Pelter, A. Reid, *Tetrahedron* **2000**, *56*, 7331–7338.

[152] C. J. Friedmann, S. Ay, S. Bräse, *J. Org. Chem.* **2010**, *75*, 4612–4614.

[153] K. Schwekendiek, F. Glorius, *Synthesis* **2006**, *2006*, 2996–3002.

[154] D. A. Horne, K. Yakushijin, G. Büchi, *Heterocycles* **1994**, *1*, 139–153.

[155] C. Uma Maheswari, G. Sathish Kumar, M. Venkateshwar, *RSC Adv.* **2014**, *4*, 39897-39900.

[156] M. Ishihara, H. Togo, *Tetrahedron* **2007**, *63*, 1474–1480.

[157] C. Sarcher, A. Lühl, F. C. Falk, S. Lebedkin, M. Kühn, C. Wang, J. Paradies, M. M. Kappes, W. Klopper, P. W. Roesky, *Eur. J. Inorg. Chem.* **2012**, *2012*, 5033–5042.

[158] J. E. Glover, P. G. Plieger, G. J. Rowlands, *Aust. J. Chem.* **2014**, *67*, 374–380.

[159] R. J. Keaton, J. M. Blacquiere, R. T. Baker, *J. Am. Chem. Soc.* **2007**, *129*, 1844–1845.

[160] M. Van Overschelde, E. Vervecken, S. G. Modha, S. Cogen, E. Van der Eycken, J. Van der Eycken, *Tetrahedron* **2009**, *65*, 6410–6415.

[161] A. Staubitz, A. P. M. Robertson, I. Manners, *Chem. Rev.* **2010**, *110*, 4079–4124.

[162] J. M. R. Narayanam, C. R. J. Stephenson, *Chem. Soc. Rev.* **2011**, *40*, 102–113.

[163] P. G. Bomben, K. C. D. Robson, P. A. Sedach, C. P. Berlinguette, *Inorg. Chem.* **2009**, *48*, 9631–9643.

[164] I. S. Kovalev, D. S. Kopchuk, G. V. Zyryanov, V. L. Rusinov, O. N. Chupakhin, V. N. Charushin, *Russ. Chem. Rev.* **2015**, *84*, 1191–1225.

[165] G. Clayden, W. Warren, *Organic Chemistry. Sl: Oxford University Press, 2001*, S, **n.d.**

[166] C. M. P. Kronenburg, J. T. B. H. Jastrzebski, J. Boersma, M. Lutz, A. L. Spek, G. van Koten, *J. Am. Chem. Soc.* **2002**, *124*, 11675–11683.

[167] D. Gelman, L. Jiang, S. L. Buchwald, *Org. Lett.* **2003**, *5*, 2315–2318.

[168] X.-L. Chen, R. Yu, X.-Y. Wu, D. Liang, J.-H. Jia, C.-Z. Lu, *Chem. Commun.* **2016**, *52*, 6288–6291.

[169] D. Hérault, D. H. Nguyen, D. Nuel, G. Buono, *Chem. Soc. Rev.* **2015**, *44*, 2508–2528.

[170] G. Laven, M. Kullberg, **2011**.

[171] Y. Li, S. Das, S. Zhou, K. Junge, M. Beller, *J. Am. Chem. Soc.* **2012**, *134*, 9727–9732.

[172] C. Petit, E. Poli, A. Favre-Réguillon, L. Khrouz, S. Denis-Quanquin, L. Bonneviot, G. Mignani, M. Lemaire, *ACS Catal.* **2013**, *3*, 1431–1438.

[173] M. Schirmer, S. Jopp, J. Holz, A. Spannenberg, T. Werner, *Adv. Synth. Catal.* **2016**, *358*, 26–29.

[174] M. E. Marmion, K. J. Takeuchi, *J. Am. Chem. Soc.* **1988**, *110*, 1472–1480.

[175] B. Matt, C. Coudret, C. Viala, D. Jouvenot, F. Loiseau, G. Izzet, A. Proust, *Inorg. Chem.* **2011**, *50*, 7761–7768.

[176] H. Schmidbaur, A. Schier, *Z. Für Naturforschung B* **2011**, *66*, 329–350.

[177] M. Brissard, O. Convert, M. Gruselle, C. Guyard-Duhayon, R. Thouvenot, *Inorg. Chem.* **2003**, *42*, 1378–1385.

[178] T. Matsui, H. Sugiyama, M. Nakai, Y. Nakabayashi, *Chem. Pharm. Bull. (Tokyo)* **2016**, *64*, 282–286.

[179] A. D. Ryabov, R. Le Lagadec, H. Estevez, R. A. Toscano, S. Hernandez, L. Alexandrova, V. S. Kurova, A. Fischer, C. Sirlin, M. Pfeffer, *Inorg. Chem.* **2005**, *44*, 1626–1634.

[180] C. Li, D. Chen, W. Tang, *Synlett* **2016**, *27*, 2183–2200.

[181] K. H. Meyer, K. Schuster, *Berichte Dtsch. Chem. Ges. B Ser.* **1922**, *55*, 819–823.

[182] M. B. Erman, I. S. Aul'chenko, L. A. Kheifits, V. G. Dulova, Ju. N. Novikov, M. E. Vol'pin, *Tetrahedron Lett.* **1976**, *17*, 2981–2984.

[183] B. M. Choudary, A. Durga Prasad, V. L. K. Valli, *Tetrahedron Lett.* **1990**, *31*, 7521–7522.

[184] P. Chabardes, *Tetrahedron Lett.* **1988**, *29*, 6253–6256.

[185] K. Narasaka, H. Kusama, Y. Hayashi, *Tetrahedron* **1992**, *48*, 2059–2068.

[186] C. Y. Lorber, J. A. Osborn, *Tetrahedron Lett.* **1996**, *37*, 853–856.

[187] T. Suzuki, M. Tokunaga, Y. Wakatsuki, *Tetrahedron Lett.* **2002**, *43*, 7531–7533.

[188] M. Picquet, A. Fernández, C. Bruneau, P. H. Dixneuf, *Eur. J. Org. Chem.* **2000**, *2000*, 2361–2366.

[189] R. S. Ramón, N. Marion, S. P. Nolan, *Recent Dev. Gold Catal.* **2009**, *65*, 1767–1773.

[190] R. S. Ramón, S. Gaillard, A. M. Z. Slawin, A. Porta, A. D'Alfonso, G. Zanoni, S. P. Nolan, *Organometallics* **2010**, *29*, 3665–3668.

[191] M. M. Hansmann, A. S. K. Hashmi, M. Lautens, *Org. Lett.* **2013**, *15*, 3226–3229.

[192] D. A. Engel, G. B. Dudley, *Org. Lett.* **2006**, *8*, 4027–4029.

[193] M. N. Pennell, M. G. Unthank, P. Turner, T. D. Sheppard, *J. Org. Chem.* **2011**, *76*, 1479–1482.

[194] D. M. Knoll, C. Zippel, Z. HASSAN, M. Nieger, P. Weis, M. Kappes, S. Bräse, *Dalton Trans.* **2019**, DOI 10.1039/C9DT04366G.

[195] E. Spuling, Synthesis of New [2.2]Paracyclophane Derivatives for Application in Material Sciences, Dissertation, KIT, **2019**.

[196] F. Vögtle, *Cyclophane Chemistry: Synthesis, Structures, and Reactions*, John Wiley & Sons Inc, **1993**.

[197] V. I. Rozenberg, E. V. Sergeeva, V. G. Kharitonov, N. V. Vorontsova, E. V. Vorontsov, V. V. Mikul'shina, *Russ. Chem. Bull.* **1994**, *43*, 1018–1023.

[198] B. Kemper, M. von Gröning, V. Lewe, D. Spitzer, T. Otremba, N. Stergiou, D. Schollmeyer, E. Schmitt, B. J. Ravoo, P. Besenius, *Chem. - Eur. J.* **2017**, *23*, 6048–6055.

[199] J. Malberg, M. Bodensteiner, D. Paul, T. Wiegand, H. Eckert, R. Wolf, *Angew. Chem. Int. Ed.* **2014**, *53*, 2771–2775.

[200] O. Morgan, S. Wang, S.-A. Bae, R. J. Morgan, A. David Baker, T. C. Strekas, R. Engel, *J. Chem. Soc. Dalton Trans.* **1997**, 3773–3776.

[201] "5050 LED Datasheet - The Ultimate Guide!," can be found under https://sirs-e.com/general/5050-led-datasheet/, **2016**.

7 Appendices

APPENDIX 1 X-RAY DATA

Data for compound **135**.

Table 1. Crystal data and structure refinement for d1909_a.

Identification code	d1909_a	
Empirical formula	C31 H29 Au Cl2 N O P	
Formula weight	730.39	
Temperature	150(2) K	
Wavelength	0.71073 Å	
Crystal system	Monoclinic	
Space group	$P2_1/c$	
Unit cell dimensions	a = 9.2320(3) Å	α= 90°.
	b = 19.8467(6) Å	β= 90.516(1)°.
	c = 15.0410(5) Å	γ = 90°.
Volume	2755.77(15) $Å^3$	
Z	4	
Density (calculated)	1.760 Mg/m^3	
Absorption coefficient	5.616 mm^{-1}	
F(000)	1432	
Crystal size	0.260 x 0.240 x 0.010 mm^3	
Theta range for data collection	1.699 to 27.619°.	
Index ranges	-12<=h<=10, -25<=k<=25, -19<=l<=15	
Reflections collected	48233	
Independent reflections	6354 [R(int) = 0.0288]	
Completeness to theta = 25.242°	100.0 %	
Absorption correction	Semi-empirical from equivalents	
Max. and min. transmission	0.7456 and 0.5743	
Refinement method	Full-matrix least-squares on F^2	
Data / restraints / parameters	6354 / 0 / 335	
Goodness-of-fit on F^2	1.047	
Final R indices [I>2sigma(I)]	R1 = 0.0202, wR2 = 0.0478	
R indices (all data)	R1 = 0.0271, wR2 = 0.0506	
Extinction coefficient	n/a	
Largest diff. peak and hole	1.010 and -0.852 e.$Å^{-3}$	

Table 2. Atomic coordinates (x 10^4) and equivalent isotropic displacement parameters ($Å^2$x 10^3) for d1909_a. U(eq) is defined as one third of the trace of the orthogonalized U^{ij} tensor.

	x	y	z	U(eq)
Au(1)	683(1)	7571(1)	4029(1)	22(1)
Cl(1)	-726(1)	8517(1)	4286(1)	22(1)
Cl(2)	5175(1)	6238(1)	10305(1)	47(1)
P(1)	2032(1)	6653(1)	3783(1)	20(1)
O(1)	4503(3)	5348(1)	8118(2)	42(1)
N(1)	2738(3)	5810(1)	8950(2)	32(1)
C(1)	2361(3)	6096(1)	4728(2)	20(1)
C(2)	1213(3)	5835(1)	5244(2)	23(1)
C(3)	1559(3)	5309(1)	5818(2)	27(1)
C(4)	2981(3)	5159(1)	6044(2)	27(1)
C(5)	4107(3)	5533(1)	5696(2)	25(1)
C(6)	3786(3)	5945(1)	4968(2)	22(1)
C(7)	-239(3)	6178(2)	5350(2)	28(1)
C(8)	-240(3)	6707(2)	6146(2)	31(1)
C(9)	1220(3)	6792(1)	6594(2)	24(1)
C(10)	1654(3)	6378(1)	7296(2)	24(1)
C(11)	3117(3)	6256(1)	7485(2)	24(1)
C(12)	4175(3)	6547(2)	6941(2)	25(1)
C(13)	3733(3)	7078(2)	6397(2)	27(1)
C(14)	2278(3)	7203(1)	6225(2)	25(1)
C(15)	5647(3)	6242(2)	6778(2)	34(1)
C(16)	5539(3)	5589(2)	6184(2)	31(1)
C(17)	3530(3)	5763(2)	8199(2)	28(1)
C(18)	3020(4)	5395(2)	9729(2)	34(1)
C(19)	3534(4)	5793(2)	10532(2)	41(1)
C(20)	3835(3)	6861(1)	3402(2)	23(1)
C(21)	4521(3)	7418(2)	3784(2)	28(1)
C(22)	5959(4)	7558(2)	3594(2)	36(1)
C(23)	6704(4)	7143(2)	3018(2)	40(1)
C(24)	6036(3)	6594(2)	2628(2)	36(1)
C(25)	4597(3)	6451(2)	2817(2)	30(1)
C(26)	1237(3)	6118(1)	2930(2)	22(1)

C(27)	1569(3)	5434(2)	2892(2)	27(1)
C(28)	1049(4)	5042(2)	2194(2)	33(1)
C(29)	153(4)	5329(2)	1555(2)	39(1)
C(30)	-203(4)	6002(2)	1596(2)	39(1)
C(31)	345(3)	6402(2)	2280(2)	29(1)

Table 3. Bond lengths [Å] and angles [°] for d1909_a.

Au(1)-P(1)	2.2387(7)
Au(1)-Cl(1)	2.3194(6)
Cl(2)-C(19)	1.789(4)
P(1)-C(20)	1.813(3)
P(1)-C(26)	1.816(3)
P(1)-C(1) 1.825(3)	
O(1)-C(17)	1.226(4)
N(1)-C(17)	1.354(4)
N(1)-C(18)	1.454(4)
N(1)-H(1N)	0.8800
C(1)-C(6) 1.394(4)	
C(1)-C(2) 1.416(4)	
C(2)-C(3) 1.391(4)	
C(2)-C(7) 1.513(4)	
C(3)-C(4) 1.386(4)	
C(3)-H(3A)	0.9500
C(4)-C(5) 1.385(4)	
C(4)-H(4A)	0.9500
C(5)-C(6) 1.396(4)	
C(5)-C(16)	1.510(4)
C(6)-H(6A)	0.9500
C(7)-C(8) 1.594(4)	
C(7)-H(7A)	0.9900
C(7)-H(7B)	0.9900
C(8)-C(9) 1.511(4)	
C(8)-H(8A)	0.9900
C(8)-H(8B)	0.9900
C(9)-C(14)	1.392(4)
C(9)-C(10)	1.394(4)
C(10)-C(11)	1.399(4)
C(10)-H(10A)	0.9500
C(11)-C(12)	1.404(4)
C(11)-C(17)	1.500(4)
C(12)-C(13)	1.394(4)
C(12)-C(15)	1.509(4)

C(13)-C(14)	1.388(4)
C(13)-H(13A)	0.9500
C(14)-H(14A)	0.9500
C(15)-C(16)	1.578(5)
C(15)-H(15A)	0.9900
C(15)-H(15B)	0.9900
C(16)-H(16A)	0.9900
C(16)-H(16B)	0.9900
C(18)-C(19)	1.515(5)
C(18)-H(18A)	0.9900
C(18)-H(18B)	0.9900
C(19)-H(19A)	0.9900
C(19)-H(19B)	0.9900
C(20)-C(25)	1.393(4)
C(20)-C(21)	1.395(4)
C(21)-C(22)	1.389(4)
C(21)-H(21A)	0.9500
C(22)-C(23)	1.384(5)
C(22)-H(22A)	0.9500
C(23)-C(24)	1.381(5)
C(23)-H(23A)	0.9500
C(24)-C(25)	1.391(4)
C(24)-H(24A)	0.9500
C(25)-H(25A)	0.9500
C(26)-C(31)	1.392(4)
C(26)-C(27)	1.392(4)
C(27)-C(28)	1.389(4)
C(27)-H(27A)	0.9500
C(28)-C(29)	1.385(5)
C(28)-H(28A)	0.9500
C(29)-C(30)	1.377(5)
C(29)-H(29A)	0.9500
C(30)-C(31)	1.391(4)
C(30)-H(30A)	0.9500
C(31)-H(31A)	0.9500
P(1)-Au(1)-Cl(1)	179.72(2)

C(20)-P(1)-C(26)	106.04(13)
C(20)-P(1)-C(1)	103.74(13)
C(26)-P(1)-C(1)	105.04(12)
C(20)-P(1)-Au(1)	112.43(10)
C(26)-P(1)-Au(1)	111.77(10)
C(1)-P(1)-Au(1)	116.89(9)
C(17)-N(1)-C(18)	122.6(3)
C(17)-N(1)-H(1N)	118.7
C(18)-N(1)-H(1N)	118.7
C(6)-C(1)-C(2)	119.2(2)
C(6)-C(1)-P(1)	118.8(2)
C(2)-C(1)-P(1)	121.9(2)
C(3)-C(2)-C(1)	116.5(3)
C(3)-C(2)-C(7)	118.0(3)
C(1)-C(2)-C(7)	124.2(3)
C(4)-C(3)-C(2)	121.7(3)
C(4)-C(3)-H(3A)	119.1
C(2)-C(3)-H(3A)	119.1
C(5)-C(4)-C(3)	120.3(3)
C(5)-C(4)-H(4A)	119.9
C(3)-C(4)-H(4A)	119.9
C(4)-C(5)-C(6)	117.1(3)
C(4)-C(5)-C(16)	120.9(3)
C(6)-C(5)-C(16)	121.1(3)
C(1)-C(6)-C(5)	121.5(3)
C(1)-C(6)-H(6A)	119.3
C(5)-C(6)-H(6A)	119.3
C(2)-C(7)-C(8)	112.5(2)
C(2)-C(7)-H(7A)	109.1
C(8)-C(7)-H(7A)	109.1
C(2)-C(7)-H(7B)	109.1
C(8)-C(7)-H(7B)	109.1
H(7A)-C(7)-H(7B)	107.8
C(9)-C(8)-C(7)	113.7(2)
C(9)-C(8)-H(8A)	108.8
C(7)-C(8)-H(8A)	108.8
C(9)-C(8)-H(8B)	108.8

C(7)-C(8)-H(8B)	108.8
H(8A)-C(8)-H(8B)	107.7
C(14)-C(9)-C(10)	116.7(3)
C(14)-C(9)-C(8)	121.0(3)
C(10)-C(9)-C(8)	121.4(3)
C(9)-C(10)-C(11)	121.8(3)
C(9)-C(10)-H(10A)	119.1
C(11)-C(10)-H(10A)	119.1
C(10)-C(11)-C(12)	119.0(3)
C(10)-C(11)-C(17)	119.8(3)
C(12)-C(11)-C(17)	120.8(3)
C(13)-C(12)-C(11)	116.9(3)
C(13)-C(12)-C(15)	117.8(3)
C(11)-C(12)-C(15)	124.1(3)
C(14)-C(13)-C(12)	121.5(3)
C(14)-C(13)-H(13A)	119.3
C(12)-C(13)-H(13A)	119.3
C(13)-C(14)-C(9)	120.2(3)
C(13)-C(14)-H(14A)	119.9
C(9)-C(14)-H(14A)	119.9
C(12)-C(15)-C(16)	111.6(2)
C(12)-C(15)-H(15A)	109.3
C(16)-C(15)-H(15A)	109.3
C(12)-C(15)-H(15B)	109.3
C(16)-C(15)-H(15B)	109.3
H(15A)-C(15)-H(15B)	108.0
C(5)-C(16)-C(15)	112.7(2)
C(5)-C(16)-H(16A)	109.0
C(15)-C(16)-H(16A)	109.0
C(5)-C(16)-H(16B)	109.0
C(15)-C(16)-H(16B)	109.0
H(16A)-C(16)-H(16B)	107.8
O(1)-C(17)-N(1)	122.2(3)
O(1)-C(17)-C(11)	123.2(3)
N(1)-C(17)-C(11)	114.6(3)
N(1)-C(18)-C(19)	113.5(3)
N(1)-C(18)-H(18A)	108.9

C(19)-C(18)-H(18A)	108.9
N(1)-C(18)-H(18B)	108.9
C(19)-C(18)-H(18B)	108.9
H(18A)-C(18)-H(18B)	107.7
C(18)-C(19)-Cl(2)	111.4(2)
C(18)-C(19)-H(19A)	109.3
Cl(2)-C(19)-H(19A)	109.3
C(18)-C(19)-H(19B)	109.3
Cl(2)-C(19)-H(19B)	109.3
H(19A)-C(19)-H(19B)	108.0
C(25)-C(20)-C(21)	119.5(3)
C(25)-C(20)-P(1)	122.5(2)
C(21)-C(20)-P(1)	117.7(2)
C(22)-C(21)-C(20)	120.4(3)
C(22)-C(21)-H(21A)	119.8
C(20)-C(21)-H(21A)	119.8
C(23)-C(22)-C(21)	119.4(3)
C(23)-C(22)-H(22A)	120.3
C(21)-C(22)-H(22A)	120.3
C(24)-C(23)-C(22)	120.9(3)
C(24)-C(23)-H(23A)	119.6
C(22)-C(23)-H(23A)	119.6
C(23)-C(24)-C(25)	119.9(3)
C(23)-C(24)-H(24A)	120.1
C(25)-C(24)-H(24A)	120.1
C(24)-C(25)-C(20)	119.9(3)
C(24)-C(25)-H(25A)	120.0
C(20)-C(25)-H(25A)	120.0
C(31)-C(26)-C(27)	119.7(3)
C(31)-C(26)-P(1)	119.5(2)
C(27)-C(26)-P(1)	120.8(2)
C(28)-C(27)-C(26)	120.2(3)
C(28)-C(27)-H(27A)	119.9
C(26)-C(27)-H(27A)	119.9
C(29)-C(28)-C(27)	119.6(3)
C(29)-C(28)-H(28A)	120.2
C(27)-C(28)-H(28A)	120.2

C(30)-C(29)-C(28)	120.6(3)
C(30)-C(29)-H(29A)	119.7
C(28)-C(29)-H(29A)	119.7
C(29)-C(30)-C(31)	120.1(3)
C(29)-C(30)-H(30A)	120.0
C(31)-C(30)-H(30A)	120.0
C(30)-C(31)-C(26)	119.8(3)
C(30)-C(31)-H(31A)	120.1
C(26)-C(31)-H(31A)	120.1

Symmetry transformations used to generate equivalent atoms:

Table 4. Anisotropic displacement parameters (Å^{2}x 10^3) for d1909_a. The anisotropic displacement factor exponent takes the form: $-2\pi^2[h^2 a^{*2}U^{11} + ... + 2 h k a^* b^* U^{12}]$

	$U^{11}U^{22}$	U^{33}	U^{23}	U^{13}	U^{12}	
Au(1)	22(1)	21(1)	24(1)	1(1)	0(1)	-1(1)
Cl(1)	22(1)	19(1)	26(1)	0(1)	1(1)	2(1)
Cl(2)	62(1)	34(1)	45(1)	3(1)	-12(1)	1(1)
P(1)	20(1)	20(1)	21(1)	0(1)	0(1)	-2(1)
O(1)	48(2)	48(2)	32(1)	6(1)	5(1)	25(1)
N(1)	32(2)	37(2)	27(1)	4(1)	1(1)	11(1)
C(1)	26(2)	17(1)	18(1)	-2(1)	1(1)	-1(1)
C(2)	25(2)	22(1)	23(1)	-4(1)	-2(1)	-5(1)
C(3)	35(2)	21(1)	25(2)	-2(1)	3(1)	-8(1)
C(4)	41(2)	18(1)	23(1)	0(1)	3(1)	5(1)
C(5)	30(2)	23(1)	23(1)	-3(1)	3(1)	10(1)
C(6)	25(2)	21(1)	21(1)	-3(1)	5(1)	2(1)
C(7)	21(2)	34(2)	28(2)	3(1)	0(1)	-7(1)
C(8)	25(2)	40(2)	28(2)	1(1)	1(1)	7(1)
C(9)	25(2)	24(1)	22(1)	-6(1)	1(1)	7(1)
C(10)	25(2)	23(1)	23(1)	-4(1)	3(1)	2(1)
C(11)	27(2)	23(1)	22(1)	-5(1)	-1(1)	4(1)
C(12)	24(2)	28(2)	24(1)	-7(1)	-2(1)	0(1)
C(13)	29(2)	26(2)	26(2)	-6(1)	1(1)	-4(1)
C(14)	34(2)	21(1)	21(1)	-4(1)	-1(1)	4(1)
C(15)	22(2)	46(2)	34(2)	1(1)	-2(1)	4(1)

C(16)	26(2)	40(2)	28(2)	4(1)	4(1)	12(1)
C(17)	30(2)	31(2)	23(1)	-1(1)	-2(1)	4(1)
C(18)	36(2)	38(2)	28(2)	8(1)	0(1)	5(1)
C(19)	48(2)	49(2)	27(2)	6(1)	3(2)	13(2)
C(20)	22(1)	26(1)	21(1)	5(1)	-2(1)	-2(1)
C(21)	26(2)	26(2)	31(2)	2(1)	-1(1)	-3(1)
C(22)	27(2)	37(2)	44(2)	3(1)	-6(1)	-10(1)
C(23)	21(2)	54(2)	44(2)	8(2)	3(1)	-4(2)
C(24)	27(2)	50(2)	32(2)	0(1)	6(1)	3(1)
C(25)	30(2)	34(2)	26(2)	-1(1)	2(1)	-4(1)
C(26)	21(1)	26(1)	19(1)	-1(1)	1(1)	-5(1)
C(27)	25(2)	26(2)	29(2)	-1(1)	0(1)	-3(1)
C(28)	36(2)	30(2)	34(2)	-7(1)	6(1)	-8(1)
C(29)	39(2)	49(2)	29(2)	-10(1)	0(1)	-15(2)
C(30)	34(2)	57(2)	26(2)	3(2)	-7(1)	-4(2)
C(31)	28(2)	32(2)	27(2)	2(1)	-1(1)	-1(1)

Table 5. Hydrogen coordinates (x 10^4) and isotropic displacement parameters (Å^{2}x 10 ^{3}) for d1909_a.

	x	y	z	U(eq)
H(1N)	2025	6104	8967	41(10)
H(3A)	800	5046	6061	32
H(4A)	3184	4797	6440	33
H(6A)	4556	6125	4628	27
H(7A)	-988	5831	5460	33
H(7B)	-497	6412	4790	33
H(8A)	-563	7150	5914	37
H(8B)	-951	6560	6596	37
H(10A)	937	6174	7656	28
H(13A)	4443	7361	6137	32
H(14A)	2004	7570	5854	30
H(15A)	6106	6126	7356	41
H(15B)	6269	6578	6480	41
H(16A)	6331	5595	5745	38

H(16B)	5673	5187	6565	38
H(18A)	3763	5055	9580	41
H(18B)	2121	5151	9885	41
H(19A)	3697	5482	11038	50
H(19B)	2774	6118	10704	50
H(21A)	4002	7702	4177	33
H(22A)	6426	7937	3856	43
H(23A)	7690	7237	2889	48
H(24A)	6558	6314	2231	44
H(25A)	4133	6075	2548	36
H(27A)	2152	5236	3344	32
H(28A)	1306	4579	2155	40
H(29A)	-219	5059	1084	47
H(30A)	-824	6193	1157	47
H(31A)	111	6868	2303	35

Table 6. Torsion angles [°] for d1909_a.

C(20)-P(1)-C(1)-C(6)	0.1(2)
C(26)-P(1)-C(1)-C(6)	111.2(2)
Au(1)-P(1)-C(1)-C(6)	-124.27(19)
C(20)-P(1)-C(1)-C(2)	177.6(2)
C(26)-P(1)-C(1)-C(2)	-71.2(2)
Au(1)-P(1)-C(1)-C(2)	53.3(2)
C(6)-C(1)-C(2)-C(3)	-15.9(4)
P(1)-C(1)-C(2)-C(3)	166.6(2)
C(6)-C(1)-C(2)-C(7)	150.9(3)
P(1)-C(1)-C(2)-C(7)	-26.7(4)
C(1)-C(2)-C(3)-C(4)	15.7(4)
C(7)-C(2)-C(3)-C(4)	-151.9(3)
C(2)-C(3)-C(4)-C(5)	-0.2(4)
C(3)-C(4)-C(5)-C(6)	-14.9(4)
C(3)-C(4)-C(5)-C(16)	154.3(3)
C(2)-C(1)-C(6)-C(5)	1.0(4)
P(1)-C(1)-C(6)-C(5)	178.6(2)
C(4)-C(5)-C(6)-C(1)	14.5(4)
C(16)-C(5)-C(6)-C(1)	-154.7(3)

C(3)-C(2)-C(7)-C(8)	79.5(3)
C(1)-C(2)-C(7)-C(8)	-87.1(3)
C(2)-C(7)-C(8)-C(9)	3.5(4)
C(7)-C(8)-C(9)-C(14)	80.8(3)
C(7)-C(8)-C(9)-C(10)	-87.7(3)
C(14)-C(9)-C(10)-C(11)	-14.9(4)
C(8)-C(9)-C(10)-C(11)	154.1(3)
C(9)-C(10)-C(11)-C(12)	-1.3(4)
C(9)-C(10)-C(11)-C(17)	-175.0(3)
C(10)-C(11)-C(12)-C(13)	16.5(4)
C(17)-C(11)-C(12)-C(13)	-169.8(3)
C(10)-C(11)-C(12)-C(15)	-151.1(3)
C(17)-C(11)-C(12)-C(15)	22.6(4)
C(11)-C(12)-C(13)-C(14)	-16.0(4)
C(15)-C(12)-C(13)-C(14)	152.4(3)
C(12)-C(13)-C(14)-C(9)	-0.4(4)
C(10)-C(9)-C(14)-C(13)	15.7(4)
C(8)-C(9)-C(14)-C(13)	-153.4(3)
C(13)-C(12)-C(15)-C(16)	-96.9(3)
C(11)-C(12)-C(15)-C(16)	70.6(4)
C(4)-C(5)-C(16)-C(15)	-95.2(3)
C(6)-C(5)-C(16)-C(15)	73.6(3)
C(12)-C(15)-C(16)-C(5)	17.0(4)
C(18)-N(1)-C(17)-O(1)	2.7(5)
C(18)-N(1)-C(17)-C(11)	-177.3(3)
C(10)-C(11)-C(17)-O(1)	137.1(3)
C(12)-C(11)-C(17)-O(1)	-36.5(4)
C(10)-C(11)-C(17)-N(1)	-42.9(4)
C(12)-C(11)-C(17)-N(1)	143.4(3)
C(17)-N(1)-C(18)-C(19)	114.3(3)
N(1)-C(18)-C(19)-Cl(2)	-59.5(3)
C(26)-P(1)-C(20)-C(25)	-25.3(3)
C(1)-P(1)-C(20)-C(25)	85.1(3)
Au(1)-P(1)-C(20)-C(25)	-147.7(2)
C(26)-P(1)-C(20)-C(21)	161.2(2)
C(1)-P(1)-C(20)-C(21)	-88.4(2)
Au(1)-P(1)-C(20)-C(21)	38.8(2)

C(25)-C(20)-C(21)-C(22)	-0.8(4)
P(1)-C(20)-C(21)-C(22)	172.9(2)
C(20)-C(21)-C(22)-C(23)	0.2(5)
C(21)-C(22)-C(23)-C(24)	0.4(5)
C(22)-C(23)-C(24)-C(25)	-0.4(5)
C(23)-C(24)-C(25)-C(20)	-0.2(5)
C(21)-C(20)-C(25)-C(24)	0.8(4)
P(1)-C(20)-C(25)-C(24)	-172.6(2)
C(20)-P(1)-C(26)-C(31)	-96.1(2)
C(1)-P(1)-C(26)-C(31)	154.4(2)
Au(1)-P(1)-C(26)-C(31)	26.7(3)
C(20)-P(1)-C(26)-C(27)	80.5(3)
C(1)-P(1)-C(26)-C(27)	-28.9(3)
Au(1)-P(1)-C(26)-C(27)	-156.6(2)
C(31)-C(26)-C(27)-C(28)	2.0(4)
P(1)-C(26)-C(27)-C(28)	-174.7(2)
C(26)-C(27)-C(28)-C(29)	-2.5(5)
C(27)-C(28)-C(29)-C(30)	1.3(5)
C(28)-C(29)-C(30)-C(31)	0.4(5)
C(29)-C(30)-C(31)-C(26)	-0.9(5)
C(27)-C(26)-C(31)-C(30)	-0.3(4)
P(1)-C(26)-C(31)-C(30)	176.4(2)

Symmetry transformations used to generate equivalent atoms:

Table 7. Hydrogen bonds for d1909_a [Å and °].

D-H...A	d(D-H)	d(H...A)	d(D...A)	<(DHA)
N(1)-H(1N)...Cl(1)#1	0.88	2.70	3.506(3)	153.7

Symmetry transformations used to generate equivalent atoms:

#1 x,-y+3/2,z+1/2

Data for compound **165/SB1259_sq**

Crystal Structure Determination of 165

The single-crystal X-ray diffraction study of **165** was carried out on a Bruker D8 Venture diffractometer with Photon detector at 123(2) K using Cu-Kα radiation (λ = 1.54178 Å. Dual space methods (SHELXT) [G. M. Sheldrick, *Acta Crystallogr.* 2015, **A71**, 3-8] were used for structure solution and refinement was carried out using SHELXL-2014 (full-matrix least-squares on F^2) [G. M. Sheldrick, *Acta Crystallogr.* 2015, **C71**, 3-8]. Hydrogen atoms were localized by difference electron density determination and refined using a riding model. A semi-empirical absorption correction was applied. The refinement with the listed atoms shows residual electron density due to a heavily disordered cyclohexane solvent molecule in one void, which could not be refined with split atoms. Therefore, the option "SQUEEZE" of the program package PLATON (A. L. Spek, *Acta Crystallogr.* 2009, **D65**, 148-155; A. L. Spek, *Acta Crystallogr.* 2015, **C71**, 9-18.) was used to create a hkl file taking into account the residual electron density in the void areas. Therefore, the atoms list and unit card do not agree (see cif-file for details).

165: colorless crystals, $C_{39}H_{32}AuClNP \cdot 0.5\ C_6H_{12}$, M_r = 820.12, crystal size 0.16 × 0.06 × 0.04 mm, triclinic, space group *P-1* (No. 2), a = 9.9785(4) Å, b = 13.0028(5) Å, c = 15.0193(5) Å, α = 114.068(1)°, β = 104.088(1)°, γ = 97.102(1)°, V = 1670.23(11) Å^3, Z = 2, ρ = 1.631 Mg/m^{-3}, μ(Cu-K$_\alpha$) = 9.70 mm^{-1}, F(000) = 816, $2\theta_{max}$ = 144.8°, 29481 reflections, of which 6556 were independent (R_{int} = 0.035), 388 parameters, R_1 = 0.029 (for 6445 $I > 2\sigma(I)$), wR_2 = 0.074 (all data), S = 1.17, largest diff. peak / hole = 2.40 (close to Au1) / -0.52 e Å^{-3}.

CCDC 1958713 (**165**) contains the supplementary crystallographic data for this paper. These data can be obtained free of charge from The Cambridge Crystallographic Data Centre via www.ccdc.cam.ac.uk/data_request/cif.

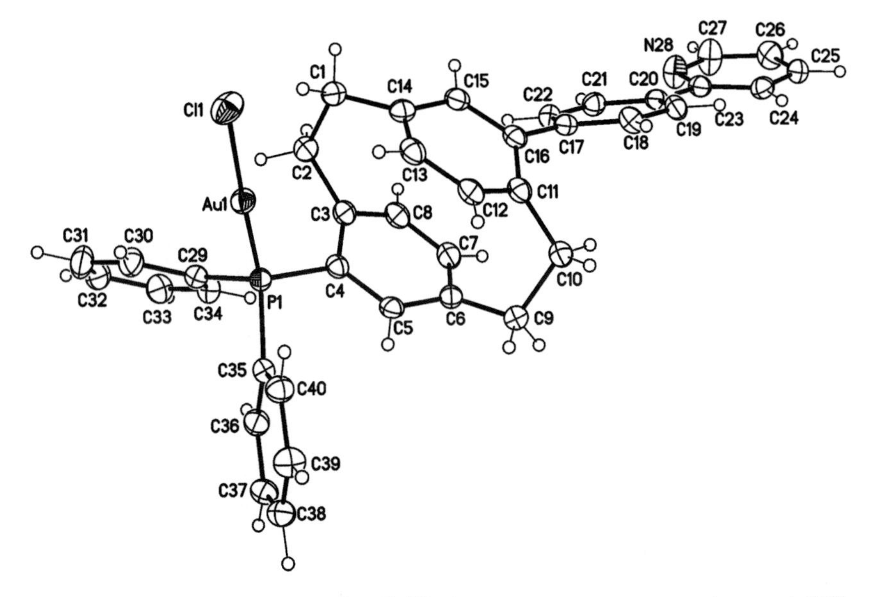

Fig. SI-1. Molecular structure of **165** (displacement parameters are drawn at 50% probability level).

APPENDIX 2 PHOTOREACTOR

The visible-light photoredox catalysis was run in a reactor built from scratch. The frame was provided by a crystallizing disk of diameter = 200 mm. The reactions were run in 5 mL glass vials with a septum crimp cap or in 1.5 mL GC vials with a septum screw-on cap. The glass vials were held by a regular lab clamp or a custom 3D-printed GC vial holder. This custom holder was kept in place by two NMR tubes glued to its bottom (Figure 7.1). To provide adequate cooling, a computer fan powered by a battery pack was placed on the bottom of the crystallizing disk, blowing room-temperature air on the reaction continuously.

The visible light is provided by a 5050 3528 RGB LED 30 LEDs/m including power supply and remote purchased on eBay. The RGB channels of this light strip can be addressed individually by the remote to make use of a specific part of the visible spectrum. Reactions marked with "green light" as reactant were run with only the green LED firing at a wavelength of 515-530 nm according to the datasheet.[201]

The advantage of this setup is the low cost (total 25€) and conveniently tunable wavelength by a remote included in the setup.

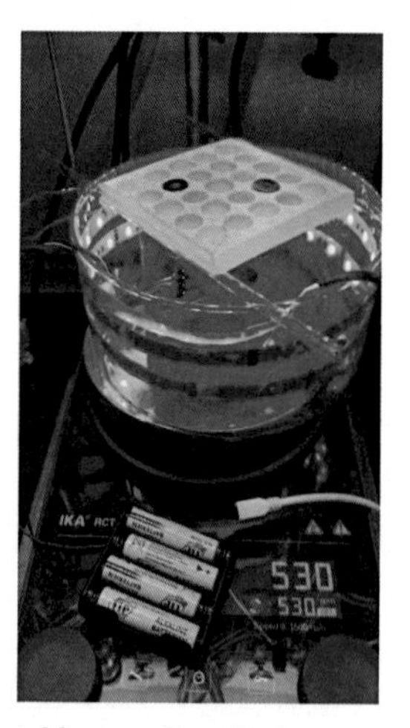

Figure 7.1. Variable wavelength photoreactor setup with 3D-printed insert for GC vials left and regular 5 mL glass vials right.

Appendix 3 List of Abbreviations and Acronyms

(v/v)	volume/volume ratio
(w/w)	weight/weight ratio
°C	degree Celsius
δ	chemical shift
µg	microgram
µL	microliter
µmol	micromole
Å	Ångström
A	Acceptor
Ac	Acyl
ACN	acetonitrile
a.q.	aqueous
Ar	aryl
ATR	attenuated total reflection
a.u.	arbitrary unit
Bu	butyl
Bz	Benzoyl
calc.	calculated
cat.	catalyst
CCDC	Cambridge Crystallographic Data Centre
CID	collision-induced dissociation
CIP	Cahn-Ingold-Prelog
CVD	chemical vapor deposition
d	day

d	doublet
D	Donor
dba	dibenzylideneacetone
DCM	dichloromethane
DDQ	2,3-dichloro-5,6-dicyano-1,4-benzoquinone
DEPT	distortionless enhancement by polarization transfer
DFT	density functional theory
DMF	*N,N*-dimethylformamide
DMSO	dimethyl sulfoxide
dppf	1,1′-bis(diphenylphosphino)ferrocene
E	Electrophile
e.g.	exempli gratia (for example)
ee	enantiomeric excess (*%ee*)
EI	electron ionization
ESI	electron spray ionization
equiv.	equivalents
Et	ethyl
et al.	*et alii* (and others)
eV	electron volt
f	oscillator strength
FAB	fast atom bombardment
FID	free induction decay
g	gram
GC	gas chromatography
GP	general procedure

h	hour
HPLC	high performance liquid chromatography
HRMS	high resolution mass spectrometry
Hz	Hertz
IR	infrared
IUPAC	International Union of Pure and Applied Chemistry
ISC	intersystem crossing
J	coupling constant
K	Kelvin
L	liter
LED	light-emitting diode
m	*meta*
M	molar
m	multiplet
m.p.	melting point
mbar	millibar
mdeg	millidegree
Me	methyl
mg	milligram
MHz	mega Hertz
min	minute
mL	milliliter
mM	milli molar
mol	millimole
MSR	Meyer-Schuster rearrangement

n-BuLi	n-butyllithium
NBS	*N*-bromosuccinimide
NMR	nuclear magnetic resonance
MS	mass spectrometry
Nu	Nucleophile
o	*ortho*
p	*para*
PCP	[2.2]paracyclophane
Ph	phenyl
ppm	parts per million
py	pyridyl
ppy	phenylpyridyl
q	quartet
R/R_P	right-handed (clockwise) stereodescriptor
r.t.	room temperature
rac	racemic
RuPhos	2-dicyclohexylphosphino-2′,6′-diisopropoxybiphenyl
s	seconds
s	singlet
S_1	first excited singlet state
S/S_P	left-handed (counter-clockwise) stereodescriptor
sat.	saturated
SPhos	2-dicyclohexylphosphino-2',6'-dimethoxybiphenyl
t	triplet
T	temperature

T_1	first excited triplet state
^{t}Bu	*tert*-butyl
THF	tetrahydrofuran
TLC	thin layer chromatography
tol	toluyl
UV	ultraviolet
V	volt
Vis	visible light
vs	very strong
vs	*versus*
vw	very weak
w	weak
W	watt
wt%	weight percent
XRC	X-ray crystallography

Appendix 4 Curriculum Vitae

Daniel Maximilian Knoll
Brieger Str. 1
Karlsruhe, 76131, Germany
Tel. (+49) 157 888 62 666
E-mail: dmk.knoll@gmail.com

EDUCATION

Ph. D., Chemistry **Anticipated graduation: December 2019**
Karlsruhe Institute of Technology (KIT), Karlsruhe
Supervision by Prof. Dr. Stefan Bräse.
Thesis Title: 'Heterobimetallic [2.2]Paracyclophane Complexes in Photoredox Catalysis'.

M. Sc., Chemistry **2014 – 2016**
Karlsruhe Institute of Technology (KIT), Karlsruhe
Supervision by Prof. Dr. Stefan Bräse.
Thesis Title: "Synthetic Approaches to Disparately Substituted [2.2]Paracyclophanes"

B. Sc., Chemistry **2011 – 2014**
Karlsruhe Institute of Technology (KIT), Karlsruhe
Supervision by Prof. Dr. Frank Breher.
Thesis Title: "Synthese und Reaktivität zweier neuer Ferrocenyl-Triazolylidliganden"

Abitur **1998 – 2011**
Eduard-Spranger-Gymnasium, Landau i.d. Pfalz

AWARDS

	Research Travel Grant, Karlsruhe House of Young Scientists (KHYS)
2018	DLR_Graduate_Program, German Aerospace Center
	NEULAND – Innovation Contest 2018, KIT
2017	Poster Award, ESOC 2017
2013	August-Wilhelm-von-Hofmann Scholarship, GDCh
2013	e-fellows Scholarship, e-fellows.net
2011	GDCh Abiturientenpreis

Appendix 5 List of Publications

Publications with peer review process

1. Knoll D. M., Zippel C., Hassan Z., Nieger M., Weis P., Kappes M. M., Bräse S., *Dalton Trans..,* **2019**, 48, 17704-17708. "A Highly Stable, Au/Ru Heterobimetallic Photoredox Catalyst with a [2.2]Paracyclophane Backbone".

2. Knoll D. M.,† Hu Y.,† Hassan Z., Bräse S., *Molecules,* **2019**, 24, 4122. "Planar Chiral [2.2]Paracyclophane-based BOX Ligands and Application in Cu-mediated N-H insertion reaction".

3. Busch J., Knoll D. M., Zippel C., Bräse S., Bizzarri C., *Dalton Trans.,* **2019**, *48,* 15338-15357. "Metal-supported and –assisted stereoselective cooperative photocatalysis".

4. Knoll D. M.,† Wiesner T. B.,† Marschner S. M., Hassan Z., Weis P., Kappes M. M., Nieger M., Bräse S., *RSC. Adv.,* **2019**, *9,* 30541-30544. "Synthesis and Characterization of Rigid [2.2]Paracyclophaneporphyrin Conjugates as Scaffolds for Fixed-distance Multimetallic Complexes".

5. Knoll D. M., Simek H., Hassan Z., Bräse S., *Eur. J. Org. Chem,* **2019**, *36,* 6198-6202. "Preparation and Synthetic Applications of [2.2]Paracyclophane Trifluoroborates: An Efficient and Convenient Route to Nucleophilic [2.2]Paracyclophane Cross-Coupling Building Blocks".

6. Hassan Z., Spuling E., Knoll D. M., Bräse S., *Angew. Chem. Int. Ed.,* **2019**, *accepted author manuscript.* "Regioselective Functionalization of [2.2]Paracyclophanes: Recent Synthetic Progress and Perspectives".

7. Knoll D. M., Bräse S., *ACS Omega,* **2019,** *3,* 12158-12162. "Suzuki Cross-Coupling of [2.2]Paracyclophane-Trifluoroborates with Pyridyl and Pyrimidyl Building Blocks".

8. Hassan Z., Spuling E., Knoll D. M., Lahann J., Bräse S., *Chem. Soc. Rev.,* **2018,** *47,* 6947-6963. "Planar Chiral [2.2]Paracyclophanes: From Synthetic Curiosity to Applications in Asymmetric Synthesis and Materials".

9. Bizzarri C., Spuling E., Knoll D. M., Volz D., Bräse S., *Coord. Chem. Rev.,* **2017**, *373,* 49-82. "Sustainable metal complexes for organic light-emitting diodes (OLEDs)".

† = equal author contribution

Publications without peer review process

1. Knoll D. M., Bräse S., Proceedings of the 43rd International Conference on Coordination Chemistry at Sendai, Japan, August **2018**. "Rigid bitopic Ligands from [2.2]Paracyclophane for Cooperative Catalysis". (Poster)

2. Knoll D. M., Spuling E., Braun C., Bräse S., Proceedings of the 20th European Symposium on Organic Chemistry at Cologne, Germany, July **2017**. "A Novel Cross-Coupling Procedure for the Convenient Heteroarylation of [2.2]Paracyclophane".

APPENDIX 6 ACKNOWLEDGEMENTS

A work like this could not have been done without the support of a great deal of people, whom I would like to thank here:

Stefan for his continued advice, supervision, numerous opportunities, giving me the freedom I needed, caring about me and late-afternoon conversations.

Prof. Lautens for the opportunity to come to Canada and learn from the best.

Prof. Roesky for being a helpful and uncomplicated correferent.

Zahid for streamlined and exponential publication growth. Make PCP great again.

Christoph Z. for proof-reading, being honest and salty, ESI adventures and listening to my endless monologues.

Christoph S. for being superspannend and approachable, arranged porphyrin-para-cyclophane marriage and reminding me what good science is.

Christian for mentoring, silent night, being outgoing and apple fritters.

Sarah for showing courage and a backbone in times of Emergency Lab Notes.

Alexander for encouraging trifluoroborates.

The groups of Prof. Kappes and Prof. Diller for fruitful collaboration.

Dr. Weis for kinetics measurements.

3MET for giving me a roof over my head and something to eat.

The research stay abroad was funded by the Karlsruhe House of Young Scientists (KHYS)
DLR for making a better a human out of me and showing me my country.

Frau Mösle, Frau Hirsch, Frau Ohmer-Scherrer and Frau Lang for their hard work and diligence.

Sophie for accepting müsli bars in the afternoon.

Jannik for showing up most of the time.

Theresa, Helena and Sven for being the hands to my brain.

Christiane, Selin and Janine for being organizational wonder workers.

Christin and Ilona for keeping the group running smoothly.

Eduard for being Eduard.

Jasmin for Japan.

Stefan M. for the way he stirred his coffee.

Nicolai for his laugh.

Thomas for invaluable life lessons, being an incredible human and careless optimism.

Fabian for 2009 until now, Yo Mama, mutual growth, Klaus Kinski, infinity minus one conversation and Äquipotentialflächen.

Anika for liking my cheese, being a partner in crime and science, believing in me and being the kindest human I know.

Mama and Papa for apparently doing more things right than I would give them credit for in raising me. Financial and moral support, tolerating my moods and accepting me.

Melanie for being an awesome sister, singing the stress away, science-code-talking our parents to insanity and sharing biology with a chemist.

Thank you.